Elham Semsar-Kazerooni • Khashayar Khorasani

Team Cooperation in a Network of Multi-Vehicle Unmanned Systems

Synthesis of Consensus Algorithms

Springer

Elham Semsar-Kazerooni
University of Toronto
Toronto, Canada

Khashayar Khorasani
Department Electrical & Computer
Engineering
Concordia University
Montreal Québec, Canada

ISBN 978-1-4614-5072-6 ISBN 978-1-4614-5073-3 (eBook)
DOI 10.1007/978-1-4614-5073-3
Springer New York Heidelberg Dordrecht London

Library of Congress Control Number: 2012950316

Printed on acid-free paper

Springer is part of Springer Science+Business Media (www.springer.com)

To My Parents, and
My Husband, Amin
Elham Semsar-Kazerooni

To My Family
Khashayar Khorasani

Preface

Recently, there has been a growing interest towards development of sensor network (SN) systems, or more broadly defined as network of unmanned systems (NUMS), that can operate autonomously for an extended period of time without an extensive involvement of human operators. This area has attracted a significant level of interest in the past few years and is currently considered as one of the strategic areas of research in the field of systems control theory. The motivation for this interest may be traced to applications where direct human intervention or involvement is either not possible or not safe due to environmental hazards, extraordinary complexity of tasks involved, or other restrictions. On the other hand, observations made based on natural behavior of animals operating as a team have inspired scientists in different disciplines to investigate the possibilities of networking a group of systems (system of systems) to accomplish a given set of tasks without requiring an explicit supervisor. Some examples of such natural behaviors can be found in the migration of birds, motion of fish searching for food, and team work of other animals which have a group living style. In all these examples, the animals work together as a team with an intergroup cooperation and with no supervision from outside the team in order to perform their complex tasks. Furthermore, it should be noted that advances in wireless communication networks have made it feasible to connect a number of systems distributed over a large geographic area.

These advances and observations, among other factors, have led the scientists to concentrate and focus on the area of network of unmanned systems. The advantages of NUMS are numerous and applications in various fields of research are being developed. Some of the advantages for deploying autonomous network of unmanned systems are enhanced group robustness to individual faults, increased and improved instrument sensing and resolution, reduced cost of operation, and adaptive reconfigurability capabilities [13]. These networks of dynamical systems are then required and tasked to cooperate in order to accomplish complicated tasks which are highly novel and challenging and which otherwise would have been impossible to accomplish by using only a single unit. These networks may be potentially constructed from integrating a large number of dynamical systems (agents), such as unmanned aerial vehicles (UAVs), unmanned ground vehicles (UGVs), unmanned underwater vehicles (UUVs), satellites [as in precision formation flight (PFF)], or mobile robots. Any one of the above systems usually consists of a number of sensors, actuators, and decision makers, and therefore the network of these systems is a network of potentially large number of sensors and actuators, or as is known in the literature, a sensor–decider–actuator network. In order to fully take advantage of these large-scale networks, several prerequisites have to be satisfied. Some of these prerequisites are development of reliable communication, optimal power consumption management, security, optimal cooperation, and team collaboration as discussed in [119].

Team cooperation and coordination for accomplishing predefined goals and requirements constitute as one of the main prerequisites for these networked multi-agent systems as they are intended to be deployed in challenging and complex missions. Although, a large body of literature has focused on team cooperation in various contexts, most of the proposed strategies are not based on rigorous synthesis approaches. This book seeks to propose and develop synthesis-based strategies for team cooperation in a network by designing optimal consensus algorithms. Towards this end, a framework for modelling and control of a network of mobile robots in a cooperative manner is developed. Specifically, two control algorithms are designed for consensus-seeking problem in such a network. First, the problem is solved by utilizing optimal control theory and by solving the corresponding Hamilton–Jacobi–Bellman (HJB) equations. The consensus problem is formally defined and solved for a team of dynamical systems having a linear dynamical model. Stability of the team is guaranteed by using modified consensus algorithms that are derived by minimizing a set of individual cost functions. An alternative approach for obtaining an optimal consensus algorithm is presented by invoking a state decomposition methodology and by transforming the consensus-seeking problem into a stabilization problem.

In the second methodology, the game theory approach is used to formulate the consensus-seeking problem in a strictly cooperative framework. For this purpose, the team cost function is defined and the corresponding min–max problem is solved in order to obtain a cooperative optimal solution. It is shown that the results obtained yield lower cost values when compared to those that are obtained by using the optimal control technique. In both approaches based on the game theory and the state decomposition technique, linear matrix inequalities are used to simultaneously impose: (a) the decentralized requirement of the solution sought and (b) the consensus constraints on the resulting designed controllers. Moreover, the performance and stability properties of the designed cooperative team are analyzed in the presence of actuator anomalies corresponding to three types of faults. Steady state behavior of the team members is also analyzed under faulty scenarios. The adaptability of the team members to unanticipated circumstances is demonstrated and verified formally. Finally, the assumption of having a fixed and undirected network topology is relaxed in order to address and solve more realistic and practical scenarios. Specifically, it is shown that stability and consensus achievement of the network subject to both random switching structures and leader assignment changes can still be achieved. Moreover, by introducing additional criteria, the desirable performance specifications of the team can still be ensured and guaranteed.

This book is mainly intended and targeted for researchers, scientists, and graduate-level university students who are interested in becoming familiar with design of consensus-seeking algorithms in the general area of networked dynamical systems. The funding for much of the research reported in this book was provided by the support from the natural sciences and engineering research council of Canada (NSERC). The second author would also like to sincerely acknowledge the funding support he has received as a Tier I Concordia University Research Chair and the support from the Faculty of Engineering and Computer Science.

Toronto, Canada E. Semsar–Kazerooni
Montreal, Canada K. Khorasani

Acknowledgments

I am very grateful to my academic advisor, Prof. K. Khorasani, for the time and effort he dedicated to this research during and after my PhD study at Concordia University. He was always a great source of inspiration, help, support, and advice for me along the challenging path of graduate studies.

I would like to sincerely thank all my friends and colleagues at Concordia University for their support in all aspects of my academic life. This specially goes to Dr. Behzad Samadi, Mr. Hani Khoshdel–Nikkhoo, Dr. Kamal Bouyoucef, Ms. Mina Yazdanpanah, Dr. Mohsen Zamani–Fekri, Dr. Nader Meskin, and Dr. Rui Ru Chen, for their kindness and very supportive attitude in the course of my PhD study.

My greatest love and gratitude is for my wonderful family members for their incredible passion, love, and dedication. First, to my husband and best friend, Amin, who provided the greatest encouragement for me to continue my work. My special gratitude goes to my father for his wisdom, for the enthusiasm he gave me for graduate studies, and for his support by every means. I am very grateful to my lovely mother for the priceless love with which she continually filled my life. Last but not least is a note of love and thanks for my beloved daughter, Darya, whose endless effort and curiosity offer me infinite power and encouragement to overcome the challenges in my academic career.

Toronto, Canada Elham Semsar–Kazerooni

Contents

1 Introduction 1
1.1 Motivation 1
1.2 Applications 2
1.3 Literature Review 3
1.3.1 Formation Control 5
1.3.2 Flocking-/Swarming-Based Approaches 9
1.3.3 Consensus Algorithms 10
1.4 General Statement of the Problem 15
1.5 Outline of the Book 17

2 Background 19
2.1 A Network of Multi-agent Systems 19
2.2 Information Structure and Neighboring Sets 20
2.2.1 Leaderless Structure 20
2.2.2 Modified Leader-Follower Structure 21
2.2.3 Ring Topology 22
2.3 Model of Interactions Among the Team Members 22
2.4 Dynamical Model of an Agent 24
2.4.1 Mobile Robot Dynamical Model: Double Integrator Dynamical Model 24
2.4.2 Linear Dynamical Model 27
2.5 Terminologies and Definitions 28
2.6 Types of Actuator Faults 30
2.7 Hamilton–Jacobi–Bellman Equations 30
2.8 LMI Formulation of the Linear Quadratic Regulator Problem 31
2.9 Cooperative Game Theory 33
2.10 Problem Statement: Consensus in a Team of Multi-agent Systems 36

3 Semi-decentralized Optimal Consensus Strategies 37
3.1 Optimal Control Problem Formulation 38
3.1.1 Definition of Cost Functions 38
3.1.2 HJB Equations for the Consensus-Seeking Problem 40
3.2 Case I: Team of Multi-agent Systems with Double Integrator Dynamical Model 43
3.2.1 Consensus Problem in a Leaderless Multi-vehicle Team 43
3.2.2 Consensus Problem in an MLF Multi-vehicle Team 45
3.3 Case II: Team of Multi-agent Systems with Linear Dynamical Model 47
3.3.1 Consensus Problem in an MLF Multi-vehicle Team 47
3.3.2 Consensus Problem in an LL Multi-vehicle Team 50
3.4 Simulation Results 52
3.4.1 Double Integrator Dynamical Model 52
3.4.2 Linear Dynamical Model 53
3.5 Conclusions 60

4 Nonideal Considerations for Semi-decentralized Optimal Team Cooperation 61
4.1 Team Behavior in Presence of Actuator Faults 61
4.1.1 Team Behavior Subject to the Loss of Effectiveness Fault in an Agent's Actuator 62
4.1.2 Team Behavior Subject to an Actuator Float Fault in an Agent 63
4.1.3 Team Behavior Subject to a Lock-in-Place Fault in an Agent 66
4.1.4 Leaderless Structure 68
4.2 Switching Network Structure 68
4.2.1 Switching Control Input and Stability Analysis 70
4.2.2 Selection Criterion for κ: Performance–Control Effort Trade-Off 74
4.3 Simulation Results 74
4.3.1 Effects of Actuator Faults on the Team Performance 75
4.3.2 Team Performance in a Switching Network Topology 78
4.4 Conclusions 84

5 Linear Matrix Inequalities in the Team Cooperation Problem 85
5.1 A Cooperative Game Theory Approach to Consensus Seeking 85
5.1.1 Problem Formulation 86
5.1.2 Solution of the Minimization Problem: An LMI Formulation 87
5.1.3 An Algorithm for Obtaining a Nash Bargaining Solution 90

5.2 An LMI Approach to the Optimal Consensus Seeking 92
5.2.1 State Decomposition 93
5.2.2 Optimal Control Design 94
5.2.3 Discussion on the Graph Connectivity 99
5.3 Simulation Results 100
5.3.1 Game Theory Approach 100
5.3.2 LMI-Based Optimal Control Approach 101
5.4 Conclusions 113

6 Conclusions and Future Work 115
6.1 Conclusions 115
6.2 Further Extensions 116

A Proofs 119
A.1 Proofs of the Lemmas and Theorems of Chap. 3 119
A.2 Proofs of the Lemmas and Theorems of Chap. 4 126
A.3 Proofs of the Lemmas and Theorems of Chap. 5 136

References 141

Index 149

List of Figures

Fig. 2.1 Information structure in (**a**) a leaderless structure and (**b**) a modified leader–follower structure ... 20
Fig. 2.2 Information structure in a ring topology ... 22
Fig. 3.1 (**a**) The x-component of the average velocity profile, (**b**) the y-component of the average velocity profile, and (**c**) the $x-y$ path trajectories of the *LL* team of four agents (Monte-Carlo simulation runs) resulting from the optimal control strategy in the infinite horizon scenario ... 54
Fig. 3.2 (**a**) The x-component of the average velocity profile, (**b**) the y-component of the average velocity profile, and (**c**) the $x-y$ path trajectories of the *LL* team of four agents (Monte-Carlo simulation runs) resulting from the optimal control strategy in the finite horizon scenario ... 55
Fig. 3.3 (**a**) The x-component of the average velocity profile, (**b**) the y-component of the average velocity profile, and (**c**) the $x-y$ path trajectories of the *MLF* team of four agents (Monte-Carlo simulation runs) resulting from the optimal control strategy in the finite horizon scenario ... 56
Fig. 3.4 (**a**) The x-component of the average velocity profile, (**b**) the y-component of the average velocity profile, and (**c**) the $x-y$ path trajectories of the *MLF* team of four agents (Monte-Carlo simulation runs) resulting from the optimal control strategy in the infinite horizon scenario ... 57
Fig. 3.5 (**a**) The x-component of the velocity profile, (**b**) the y-component of the velocity profile, and (**c**) The $x-y$ path trajectories of the *MLF* team of four agents with linear dynamical model resulting from the optimal control strategy in the infinite horizon scenario ... 58

Fig. 3.6 (**a**) The x-component of the velocity profile, (**b**) the y-component of the velocity profile, and (**c**) the $x-y$ path trajectories of the *LL* team of four agents with linear dynamical model resulting from the optimal control strategy in an infinite horizon scenario 59

Fig. 4.1 Switching signal $\sigma(t)$ 70

Fig. 4.2 (**a**) The x- and (**b**) the y-components of the velocity profiles and (**c**) the x-y path trajectories of the *MLF* team of four agents in presence of the *LOE* fault in the fourth vehicle during $115 \leq t \leq 135$, where $u_f^4 = 0.5u^4$ 76

Fig. 4.3 (**a**) The x-component, (**b**) the y-component, and (**c**) the x-y path trajectories of the *MLF* team of four agents (Monte Carlo simulation runs) in the presence of a float fault in the third vehicle (velocity is frozen at $v^3 = [6\ 1]^T$). The jump in the velocity of agent 3 at $t = 30$ s is due to the initiation of a recovery procedure in the actuator of agent 3 (following the fault that is injected at $t = 20$ s) to the healthy velocity after $t \geq 30$ s 77

Fig. 4.4 (**a**) The x-component of the average velocity profiles, (**b**) the y-component of the average velocity profiles, and (**c**) the x-y path trajectories of the *MLF* team of four agents (Monte Carlo simulation runs) in presence of the float fault in the third vehicle (velocity is frozen at $v^3 = [0\ 3]^T$ during $20 \leq t \leq 30$) 79

Fig. 4.5 (**a**) The x-component of the velocity profiles, (**b**) the y-component of the velocity profiles, and (**c**) the x-y path trajectories for the *MLF* team of four agents in presence of the *LIP* fault in the third vehicle during $20.5 \leq t \leq 25$ where $u_f^3 = u^3(t = 20.5)$ 80

Fig. 4.6 (**a**) The x-y path trajectories, (**b**) the x component of the velocity profiles, and (**c**) the y component of the velocity profiles of the *MLF* team of four agents with a linear model in presence of the *LIP* fault in the third vehicle during $20.5 \leq t \leq 25$ where $u_f^3 = u^3(t = 20.5)$ 81

Fig. 4.7 The structure associated with the dynamic transitions of the team for three different switching topologies and leader assignments.. 82

Fig. 4.8 (**a**) The x-component and (**b**) the y-component of the velocity profiles and (**c**) the x-y path trajectories of the *MLF* team of four agents with switching structure and switching leader that are obtained by the application of our proposed switching control strategy 83

Fig. 5.1 (**a**) The x-component and (**b**) the y-component of the average velocity profiles that are obtained by applying the semi-decentralized optimal control strategy to a team of four agents 102
Fig. 5.2 (**a**) The x-component and (**b**) the y-component of the average velocity profiles that are obtained by applying the cooperative game theory strategy to a team of four agents..... 103
Fig. 5.3 The x-component of the average control efforts that are obtained by applying (**a**) the semi-decentralized optimal control strategy and (**b**) the cooperative game theory approach to a team of four agents 104
Fig. 5.4 (**a**) The x-component and (**b**) the y-component of the velocity profiles; optimal design based on the solution of the *LMI*s when $AS \neq 0$ in Example 1 105
Fig. 5.5 (**a**) The x-component and (**b**) the y-component of the velocity profiles; optimal design based on the solution of the Riccati equation when $AS \neq 0$ in Example 1 106
Fig. 5.6 (**a**) The x-component and (**b**) the y-component of the velocity profiles; optimal design based on the solution of the *LMI*s when $AS = 0$ in Example 1 107
Fig. 5.7 (**a**) The x-component and (**b**) the y-component of the velocity profiles; optimal design based on the solution of the Riccati equation when $AS = 0$ in Example 1 108
Fig. 5.8 Graph describing the topology of a network of multi-agent systems 109
Fig. 5.9 (**a**) The x-component and (**b**) the y-component of the velocity profiles corresponding to an optimal control design based on the solution of the *LMI*s when $AS \neq 0$ in Example 2 109
Fig. 5.10 (**a**) The x-component and (**b**) the y-component of the velocity profiles corresponding to an optimal control design based on the solution of the Riccati equation when $AS \neq 0$ in Example 2 110
Fig. 5.11 (**a**) The x-component and (**b**) the y-component of the velocity profiles corresponding to an optimal control design based on the solution of the *LMI*s when $AS = 0$ in Example 2 111
Fig. 5.12 (**a**) The x-component and (**b**) the y-component of the velocity profiles corresponding to an optimal control design based on the solution of the Riccati equation when $AS = 0$ in Example 2 112

List of Tables

Table 3.1 The mean performance index corresponding to different team structures and control design assumptions 53

Table 5.1 A comparative evaluation of the average values of the performance index corresponding to the two control design strategies for the cost functions defined in (3.5) for $T = 2\,\mathrm{s}$ 101

Table 5.2 A comparative evaluation of the performance index corresponding to the two control design strategies for the cost function (5.40) for $T = 2\,\mathrm{s}$ in Example 2 112

Table 5.3 A comparative study of the consensus-reaching time corresponding to the two control design strategies in Example 2 113

Acronyms

ARE	Algebraic Riccati equation
AUV	Autonomous underwater vehicles
BDF	Backward differentiation formula
DRE	Differential Riccati equation
GNC	Guidance, navigation, and control
HJB	Hamilton–Jacobi–Bellman
ISR	Intelligence, surveillance, and reconnaissance
LIP	Lock-in-place
LL	Leaderless
LMI	Linear matrix inequality
LOE	Loss of effectiveness
LQR	Linear quadratic regulator
MIMO	Multi-input multi-output
MLF	Modified leader–follower
NBS	Nash bargaining solution
ND	Negative definite
NSD	Negative semi-definite
NUMS	Network of unmanned systems
PDE	Partial differential equation
PD	Positive definite
PFF	Precision formation flight
PSD	Positive semi-definite
SN	Sensor networks
UAVs	Unmanned aerial vehicles
UGVs	Unmanned ground vehicles
UUVs	Unmanned underwater vehicles

Chapter 1
Introduction

Recently, there has been a growing interest towards development of sensor network (SN) systems, or more broadly defined as a network of unmanned systems (NUMS), that can operate autonomously for an extended period of time without an extensive involvement of human operators. In this chapter, we provide a brief introduction to these networks, their applications, the control problems that arise in this domain, as well as solutions that are proposed for these problems in the literature.

1.1 Motivation

Design and analysis of NUMS has attracted a significant level of attention in the past few years and is currently one of the strategic areas of research in systems theory. The motivation for this interest may be traced to emergence of applications where direct human intervention is not possible due to either environmental hazards, extraordinary complexity of the tasks, or other restrictions. On the other hand, observations made based on natural behavior of animals operating as a team have inspired scientists in different disciplines to investigate the possibilities of networking a group of systems to accomplish a given set of tasks without requiring an explicit supervisor or operator. Some examples of such natural behaviors can be found in the migration of birds, motion of fish searching for food, and team work of other animals which have a group living style. In all these examples, the animals work together in a team with an intergroup cooperation and with no supervision from outside the team in order to perform complex tasks. Furthermore, advances in wireless communication networks have made it feasible to connect a number of systems distributed over a large geographic area.

These advances and observations have led the scientists and engineers to focus and concentrate on design requirements of NUMS. The advantages of NUMS are numerous and applications in various fields of research are being developed. Some of the advantages for deploying autonomous NUMS are enhanced group robustness to individual faults, increased and improved instrument sensing and

E. Semsar-Kazerooni and K. Khorasani, *Team Cooperation in a Network of Multi-Vehicle Unmanned Systems: Synthesis of Consensus Algorithms*, DOI 10.1007/978-1-4614-5073-3_1, © Springer Science+Business Media New York 2013

resolution, reduced cost of operation, and adaptive reconfigurability capabilities [13]. In essence, these networks comprising of multi-agent systems can cooperate and coordinate their activities in order to accomplish a complicated task which is highly difficult, if not impossible, to be accomplished by using only a single agent.

These networks may be potentially constructed from a large number of dynamical systems (agents), such as satellites [as in precision formation flight (PFF)], mobile robots, unmanned aerial vehicles (UAVs), unmanned ground vehicles (UGVs), or unmanned underwater vehicles (UUVs). Associated with any realistic network of these multi-agent systems one could envisage dealing with a large number of sensors, actuators, and decision makers. In other words, one is dealing with a network of potentially thousands of sensors, actuators, and deciders, or as is often known in the literature as a sensor-decider-actuator network.

Motivated from the above discussion and given the multidisciplinary nature of the problem, it is not surprising that design of networks of sensors, actuators, and decision makers is currently one of the most important and challenging areas of research in disciplines such as communications, control theory, and mechanics. In order to fully take advantage of these large-scale networks, several prerequisites have to be satisfied. Some of these prerequisites are advances yet to be made in the development of reliable communication, optimal power consumption management, security, optimal cooperation, and team collaboration technologies as discussed in [119]. Team cooperation and coordination for accomplishing predefined goals and requirements constitute as one of the main prerequisites for these networked multi-agent systems given that they are intended to be deployed in challenging environments and missions. Although, a large body of work has been devoted to the design requirements for NUMS, unsolved problems still prevail in the literature. Some of these challenges are compounded due to (a) lack of complete information for all the agents in the network, (b) cohesion requirement and connectivity of the team in the presence of partial information, (c) uncertainties and inaccurate information due to the large-scale nature of the network, (d) presence of adversarial and environmental uncertainties, (e) robustness issues, and (f) fault diagnosis and recovery requirements, to name a few. Other issues that deserve to be addressed are the dynamic nature of the team and the use of more realistic agent dynamical models as opposed to point-mass models that are commonly used in the literature. Consequently, it is imperative that one designs reliable and high-performance NUMS that can ensure and accommodate the above requirements.

1.2 Applications

Networks of unmanned systems provide significant capabilities and functionalities and as a result have received extensive attention in the past few years where numerous applications in various fields of research are being considered and developed for them. Some applications that necessitate development of these systems are in space explorations; satellite deployment for distributed deep space observations; automated factories; guidance, navigation, and control (GNC) of a team of UAVs; utilization of a network of mobile robots for search and rescue missions; and teams of robots deployed in hazardous environments where human involvement is impossible or dangerous. In [119], other applications are provided, such as home and building automation, intelligent transportation systems, health monitoring and assisting, and commercial applications. There are also military applications in intelligence, surveillance, and reconnaissance (ISR) missions in the presence of environmental disturbances, vehicle failures, and in battlefields subject to unanticipated uncertainties and adversarial actions [16].

1.3 Literature Review

Cooperation in a NUMS, known as formation, network agreement, collective behavior, flocking, consensus, or swarming in different contexts, has received extensive attention in the past several years. Several approaches to this problem have been investigated under different frameworks and by considering different architectures [19, 32, 33, 74, 77–80, 88, 92, 94, 96, 101, 128, 141]. Moreover, the problem of cooperation in a network has been considered at various levels. At the high level, one can refer to task assignment, timing and scheduling, navigation and path planning, reconnaissance, and map building [2, 47, 69], to name a few. In the mid level, cooperative rendezvous, formation keeping, application of consensus algorithms, collective motion, and formal methods based on flocking/swarming concepts can be mentioned [13, 23, 32, 33, 79, 80, 132].

Since this book is aimed at cooperation in the mid level, therefore most of the literature that is reviewed in this section will be on cooperation at this level, i.e., formation keeping, consensus algorithms, and formal methods based on flocking/swarming concepts. The formation control problems can be distinguished from the consensus-seeking and flocking-/swarming-based approaches based on the degree of autonomy as well as on the degree of distribution of the designed algorithms. Generally, these two problems are more explicitly characterized as follows:

- *Formation Control:* Based on the definition that is presented in the survey paper by Scharf et al. [99], the formation control couples the dynamics of each member of the team through a common control law. Basically, the formation control should have two properties, namely (1) at least one member of the team must track a predefined state relative to another member and (2) the corresponding control law should be dependent on the state of this member [99]. Some of the characteristics of this type of problem are as follows:

  1. The formation control is usually solved based on a *centralized* approach to the problem [51], although in some cases *decentralized* solutions are suggested, e.g., [91]. However, the assumption of modular architecture is common in formation problem [12, 13, 31].
  2. The *number of vehicles* is usually quite limited, i.e., 5–6 vehicles [120].

3. Usually *conventional control methods* are used as in the adaptive and nonlinear control techniques [51].
4. *Information structure* is not highlighted and not embedded within the design procedure.
5. Autonomy is up to the level where a tracking path is given, a leader guides the group, or a coordination vector is provided [51].
6. Different structures, such as virtual structure [51] and Multi-input multi-output [36] architectures to the problem of formation, to name a couple, have been investigated and developed for *complex dynamics* such as aircraft, UAVs, and spacecraft.
7. Various works have covered different aspects of the control problem in this area such as formation keeping, tracking, formation stability (e.g., input-output stability), and fuel optimality [13, 100], as well as high-level tasks, such as initialization and reconfiguration [37, 63].
8. The *present challenges* are on experimental and practical issues such as design of high-resolution space instruments, fuel optimality, and reducing the computational complexity of the proposed algorithms. Also, the estimation problem in formation keeping [122, 123, 125] and design of decentralized and distributed algorithms constitute as critical and timely trends of research.

- *Flocking/Consensus/Agreement:* The problem of flocking or consensus is to have network agreement on a scalar state or a vector of states or on a function of states, while other behaviors (e.g., formation) are guaranteed. Some of the characteristics of these types of problems are as follows:

  1. They are based on a *distributed* approach and a large number of agents can be addressed [71].
  2. *Information* is considered as one part of the modelling formulation with the corresponding impact and influence of the communication topology on the stability and other dynamical properties of the network [30].
  3. The mathematical tool for the problem formulation in majority of the work is *graph theory* except for a few references, such as in [7, 24, 44, 82, 126–128].
  4. *Very simple agents' dynamical models* (first- or second-order integrator models) are mostly assumed except in few works as in [7, 30, 44, 82, 86, 96].
  5. Usually there is an *analysis* of the response and not a *synthesis*, except in few works as in [7, 24, 44, 82].
  6. Not much *experimental considerations* are devoted in the literature.
  7. Different aspects of the problem that have already been considered in the literature are as follows:

     – Basic properties such as convergence [78], finding equilibrium state [46], and notion of controllability [83] have been introduced and investigated.
     – Behavior such as formation keeping, collision avoidance, obstacle avoidance, as well as generating feasible planar trajectories to obtain the maximum coverage of the region, and determining the minimal information structure needed for stability of a swarm are addressed in [43, 73, 78, 127].

   - In [142], the main purpose is to determine the consensus algorithm weighting such that the fastest convergence speed is achieved.

8. Some of the main *challenges* presently considered as open problems in this area can be listed as follows:

   - Defining the *basic properties*, such as group stability, controllability, and observability conditions
   - Extending the existing methods to conditions where *more realistic agents' dynamical models* are considered
   - Adding conditions such as *uncertainty* in the mission plan (leader command) for the followers, model uncertainties, *disturbances*, communication/sensor noise, presence of faults in the leader and the followers
   - Putting *constraints* on the input signals and other practical constraints
   - Assuming a time-varying neighboring set, *switching* structures, and *dynamic* network topologies
   - Formulating the *estimation* problem in the network framework
   - Assuming *stochastic frameworks* with probabilistic information links
   - Considering *delayed information* exchanges among the leader and the followers or among the followers

In the following sections, we will present a detailed literature review on different issues that arise in the cooperative control.

1.3.1 Formation Control

According to its basic definition, the main focus in the formation control is for a team to achieve a predefined and given geometry and shape. This shape (formation) should be preserved during the execution of a given mission so that the team should act as a rigid body. Based on this property, a predefined trajectory is usually provided for the team motion, e.g., a leader command or a trajectory for the virtual structure center of mass. The team should track this trajectory while its members keep their relative positions and preserve the required shape, i.e., the stability of the formation should be maintained. Other requirements can be augmented to these objectives as well. A main prerequisite for maintaining the formation is guidance and control of the vehicles, while they perform tasks such as initialization, contraction, and expansion.

As discussed in [99], various architectures can be considered for the formation of a team of multi-agent systems, namely (i) treating the formation as a single multi-input multi-output (MIMO) system [36], (ii) leader-follower [118], (iii) virtual structure [13] (or virtual leader [55]), in which the entire formation is considered as a virtual structure, (iv) cyclic with nonhierarchical control architecture [139], and (v) behavioral [8].

In the MIMO architecture, the entire dynamics of the system is considered as a single MIMO model. Hence, in this architecture, any of the conventional control strategies, such as optimal, nonlinear, and robust control strategies, can be applied to the system.

The leader-follower architecture has been used very often in the literature [16,46,64,118,133,134,140]. In this approach, a hierarchical control architecture is considered with one or more of the agents as the leader(s) and other agents as the followers. The followers should track the position and orientation of the leader(s). This structure can also be constructed in a tree configuration, in which an agent is the leader of some other agents who are themselves the leaders of other agents and so on. The advantages of this approach are that it has an easy and understandable behavior, that the formation is preserved even if the leader is perturbed, and that group behavior can be inspected by defining the behavior of the leader. However, lack of an explicit feedback to the formation, from the followers to the leader, is a major disadvantage of this structure. Also, the failure of the leader implies the failure of the formation as indicated in [13,92].

In the virtual structure approach, the entire formation is treated as a single unit. In this approach, three steps are considered for the control design, namely (i) first, the desired dynamics of the virtual structure is defined, (ii) second, the states of the virtual structure is transformed into the states of individual agents, and (iii) third, the control law for each agent is designed correspondingly. In [13,92], the advantages of this method are highlighted as its simplicity in defining the coordinated behavior of the group, in keeping the formation during various maneuvers, and in existence of feedback from the agents to the virtual structure. The weakness of this structure is in its limitation in applications to time-varying or frequently reconfigurable formations. In [12, 13, 92] the idea of adding a feedback from the vehicles to the coordination unit is presented. Authors in [92] proposed virtual structure as a solution to the problem of multiple-spacecraft formation. In their approach, a feedback from a vehicle to the virtual structure is considered for keeping the vehicles in the formation and for improving the robustness of the formation in case of disturbances or when the motion of the virtual structure is rapidly changing so that some of the slow members could possibly get "lost" from the team.

In the cyclic architecture, the agents are connected to each other in a cyclic form rather than a hierarchical architecture [139]. The disadvantage of this method is that the stability analysis of the proposed control system is not straightforward due to the dependency of an individual controller on the others' in a cyclic form.

In the behavioral approach, several commands and requirements are combined or mixed together to reach various and probably competing goals or several behaviors, e.g., collision avoidance, obstacle avoidance, and formation keeping of the agents. The control law for each agent is a weighted average of the control law associated with each behavior. Since competing behaviors are averaged, occasionally strange and unpredictable behavior may occur in the overall network. Despite the advantages of simple derivation of the control strategies, the presence of an explicit feedback to the formation, and the capability of a decentralized implementation, nevertheless there are certain weaknesses as well. For example, group behavior cannot be explicitly defined and the mathematical analysis, such as stability analysis, is not straightforward as mentioned in [13].

Although the above structures are originally introduced for formation keeping, some of them can be used for flocking and consensus-based algorithms as well. Due to the supervised nature of the leader-follower structure and the virtual structure, they are not commonly used for flocking and consensus seeking, which are autonomous in their very nature (there is no predefined path, trajectory for the leader, or virtual coordinates to be tracked). However, these structures can also be redefined for utilization in flocking/swarming or consensus-seeking problems, see, e.g., [46, 140].

One of the main challenges that arise in the development of a formation keeping strategy in a NUMS is the lack of complete information and the presence of uncertainties, faults, and unpredictable events in the team. This has necessitated the design and application of adaptive methods for formation control in some applications [16, 51, 101, 118, 144]. In [16], the leader commands are unknown to the follower vehicles in a leader-follower architecture, and therefore an adaptive controller is used for the formation keeping. Missing leader commands may occur during a mission when the leader spots an approaching threat and quickly reacts to avoid it. Under this scenario there is no adequate time for the leader to communicate its new commands to the follower vehicles, and hence the leader commands need to be assumed to be unknown in the control design. Also, in [118] the authors considered uncertainties in the vehicle dynamics, in which vortex forces are considered as unknown functions. In a formation mode, each vehicle experiences an upwash field generated by the other vehicles, and therefore the aircraft motion is affected by the vortex of the adjacent vehicles. These effects are usually unknown and depend on the area, gross mass, span, and dynamic pressure characteristics as well as the velocity and position of the vehicle.

The ideas in [16, 118] are integrated in [101] by using the framework that is introduced in [16] for design of an adaptive controller in order to compensate for the time-varying unknown leader commands and vortex forces. The idea in [16] was extended to the case where the vortex forces are presented in the dynamical model of the system and both the vortex forces of the follower and the leader commands are treated as time-varying unknown parameters. The objective is to design the follower control such that the relative distances between the followers and the leader are maintained close to their desired values in presence of these uncertainties. When the vortex forces are considered in the velocity dynamics, adaptive updating laws are introduced for two cases of time-varying and constant forces. In this case the stability and tracking of relative distances are guaranteed. If in addition, the forces are applied to the heading angle dynamics, stability and tracking of the relative distances can also be guaranteed corresponding to the constant vortex forces. The proposed algorithm was applied to the formation control of UAVs.

Some of the research work conducted in the coordination problem in a team of vehicles assume graph theory as the mathematical framework for modelling a distributed network of agents [30, 73–75]. Certain approaches provide the analysis by using graph properties and this framework. The main difference between the previously reviewed work on formation control and the above references is that in the latter, the final goal is to achieve a stabilized formation in an autonomous manner. No path or trajectory is provided for the group motion and the formation is stabilized based on the intergroup exchanges of information and the desired shape that is provided a priori. In other works, the graph theory tool is used only for modelling and other tools are used for the mathematical analysis.

In [73], a local distributed bounded control input is designed for formation stabilization in a team of multiple autonomous vehicles. Point-mass dynamical models are assumed for the agents and the information flow graph can be either directed or undirected. The concept is to use potential functions which are constructed by using desired properties, such as collision-free and stable formation. Similarly, results on global stabilization and tracking are presented in [74] for a team of multi-agent systems with linear dynamics. The problem is first solved for two agents and then extended to the general case by using "dynamic node augmentation." The collision avoidance property is also guaranteed. In this approach the problems of stabilization and tracking control are decoupled into three subproblems, namely control of shape dynamics, rotational dynamics, and translational dynamics. In [75], the same problem is considered where different maneuvers such as split, rejoin, and reconfiguration are considered for the team. In [30], graph theory is used to model the communication network and to find the relationship between the topology and the stability of the formation. Based on the graph Laplacian matrix properties, a Nyquist criterion is obtained to guarantee formation stabilization for a team of multi-agent systems with linear dynamics. The formation stability is divided into two parts, namely stabilization of the information flow and the stabilization of the individual vehicles. The leader-follower architecture can be addressed in this framework.

In [128], the formation and alignment objectives are translated into an error framework. A decentralized robust controller is then designed for the error dynamics based on an overlapping design methodology. The assumed structure is leader-follower and a constant velocity command and formation structure is provided for the entire team. In [51], an adaptive approach is proposed to achieve formation of three satellites which in turn are used as a free-flying interferometer. The objective is formation of the satellites to follow a desired attitude trajectory. In [50], formation flight control of a network of UAVs is addressed.

There are other important issues that have been discussed for the formation of multi-agent systems in the literature. In the work of Smith et al. [122], a parallel estimation structure based on the error covariance is suggested for control of a formation of multi-agent systems; however the solution is not necessarily an optimal one. In their approach, each vehicle with a linear dynamical model estimates the state vector of the entire group based on the noisy output it receives. Subsequently, in the design of the controller, each vehicle uses its own state estimates. The authors have also investigated results on adding communication in order to remove the disagreement dynamics in [123, 124]. For this purpose, they have assumed that some of the nodes (at least $N-1$ receivers) send their estimation results to the rest of the team. In [121, 125], a similar idea is used for applications in spacecraft formation. In [98], convex optimization is used to develop a framework for distributed estimation problem, or equivalently for the data fusion problem. For solving the corresponding optimization problem, sub-gradient method and dual decomposition techniques are utilized.

1.3.2 Flocking-/Swarming-Based Approaches

In [67], the definition of flocking in a group of agents is given as

> A group of mobile agents is said to (asymptotically) flock, when all agents attain the same velocity vectors, and furthermore distances among the agents are asymptotically stabilized to constant values.

In [78], a dynamic graph theoretic framework is presented for formalizing the problem of flocking in presence of obstacles that are assumed to be in convex and compact sets and their boundaries are closed differentiable Jordan curves. An energy function is constructed for the team of multi-agent systems in which different tasks of flocking are considered. Dissipation of this energy function through the protocols that are inspired by the Reynolds rules [97], namely (1) alignment, (2) flock centering, and (3) collision avoidance, results in achieving all the predefined goals of the team mission. The main contribution of [78] is to derive and analyze an advanced form of Reynolds rules, specifically associated with the last two rules. The authors have considered a point-mass dynamics for the agents and a flocking protocol is defined for interactions among the agents which results in reduction of the constructed potential function. This protocol yields an alignment, i.e., convergence of each agent to the weighted average position of its neighbors, while ensuring obstacle avoidance of the network agents. Two tasks of split/rejoin and squeezing were presented in the simulation results. Similarly, in [71], a particle-based approach to flocking is considered for two scenarios, namely those associated with the presence of multiple obstacles and a free space situation. The suggested algorithms have addressed the three Reynolds rules.

In [130–132] a stable flocking motion law is introduced for a group of mobile agents with a connected graph. This law guarantees a collision-free and cohesive motion and an alignment in the headings of the agents. The authors proposed a two-component control law in which the first part is selected by using a potential function that regulates the relative distances among the agents in order to avoid collision while the second part regulates the velocities. The potential function can define the final shape of the formation. In the case of a switching topology [131, 132], control laws may be switching and therefore the Filippov and non-smooth system framework is used for stability analysis. In this case, the neighboring set is based on the distances among the agents, and therefore it is dynamic. In [53, 54], the flocking problem is solved by decomposing the entire team dynamics into the dynamics of the group formation and dynamics of the corresponding motion of the center-of-mass. Each of these dynamics is analyzed and stabilized separately such that both formation keeping and velocity regulation to a constant value are guaranteed.

1.3.3 Consensus Algorithms

Investigation on the effects of information structure on a control decision was initiated first by the work in [61] on theory of teams and later followed by [39, 40, 62, 81, 145]. These works can be considered as among the first in which the team control problem is discussed where only partial information is assumed to be available to the members. However, since these early works, in recent years a large body of research has been conducted in which the effects of information on the control design problems are discussed, see, e.g., [30,46,72,77–79,87,93,94,96,141]. In most of these works, each team member has access to limited information from other agents, or to only information of its neighbors. Furthermore, the final state of the team is to be decided by the entire team members.

Consensus algorithms are among the tools that are used for analysis of distributed systems where the network information structure has not only a crucial effect on the control design but also only certain information is available to each member. As mentioned in [79], consensus problems deal with the agreement of a group of agents upon specific "quantities of interest." In this configuration, the agents attempt to decide and agree upon among themselves what the final state should be. The state where all the quantities of interest are the same is called the consensus state. In this book, we are particularly interested in problems in which the quantities of interest are related to the motion of agents, e.g., their velocities or positions. However, as indicated in [27], several other applications for consensus algorithms exist such as those applications in decentralized computation, clock synchronization, sensor networks, and coordinated control of multi-agent networks.

In [79], linear and nonlinear consensus protocols are applied to directed and undirected networks with fixed and switching topologies. A disagreement function was introduced as a Lyapunov function to provide a tool for convergence analysis of an agreement protocol in a switching network topology. The authors have shown that the maximum time-delay that can be tolerated by a network of integrators that use a linear consensus protocol is inversely proportional to the largest eigenvalue of the Laplacian of the information flow graph or the maximum degree of the nodes of the network. Similar results are obtained in [77] where convergence analysis is developed by using Nyquist plots for linear protocols. For nonlinear protocols, the notion of action graphs and disagreement cost are introduced and the problem was solved in a distributed manner.

In [46], the coordination problem is discussed for a team of multi-agent systems using nearest neighbor rule for both the leaderless and the leader-follower configurations. The main focus of this work is on heading angle alignment in undirected graphs where the agents have simple integrator dynamics and same speed but different headings. In the leader-follower case, the leader can affect the followers whenever it is in their neighboring set. However, there is no feedback from the followers to the leader. It is shown that the connectivity of the graph on average (connection of union of graphs) is sufficient for convergence of the agent heading angles. The neighboring set assignment is switching, and therefore the team structure is dynamic. In [28], asynchronous protocols for consensus seeking are introduced. Some updating rules for the agent's control inputs with discrete-time dynamical equations are suggested so that the consensus state would take a desirable predefined value.

In [7], passivity is used to achieve network agreement (or consensus) for a class of multi-agent systems with dynamics which can satisfy the passivity conditions. The team main goal is to reach at a predefined common velocity (or any other interpretation of the derivative of a state), while the relative positions (the differences between a common state in the team) converge to a desired compact set. According to this method a Lyapunov function can be constructed for stability analysis in a distributed communication network with bidirectional links. The designed controller using passivity properties is effectively a filter which has a nonlinear function of the relative states as its input. The relationship between the topology and the stability of the formation is also provided. In [38], a wider class of systems, namely nonlinear dissipative systems, are considered and synchronization in a strongly connected network of multi-agent systems with this dynamical property is discussed.

In [83–85], an interpretation of the controllability is defined and shown for a first-order integrator model. The main purpose in these works is to find the effects of external decisions on the agreement dynamics, in particular the conditions where some of the nodes do not follow the agreement protocols (decisions). In other words, there are certain nodes that follow the agreement protocol while others have external inputs. The authors attempt to answer the question whether these "anchored" nodes are able to guide the rest of the group to the desired point. Similar to the ideas presented in [83] for a fixed network topology, the controllability conditions for a network with fixed and switching topologies are discussed in [57]. The authors have considered a leader-follower structure with one-way links from the leader to the followers. They have shown that controllability of the team is highly dependent on how the followers are connected to the leader.

The objective in [127] is to find the minimum communication that is required for guaranteeing the stability of a swarm of vehicles. The approach involves first defining a centralized cost function for the team and then dividing it into certain individual costs. A vector of parameters are introduced for quantization of the information which should be exchanged among the agents in order to achieve a stable formation and the optimization is accomplished with respect to these parameters. Similar approach is pursued more recently in [126].

In [142], a fixed and a given network structure is assumed and the question addressed is how to find the weights of the interconnection links such that the convergence to consensus value is achieved at the fastest rate. To solve this problem a set of criteria to be minimized is introduced. The resulting optimization problem is non-convex, which is then converted into a convex one. In [5], an estimate of the convergence rate of the consensus seeking is obtained. The communication links are assumed to be time-varying. In [21], a lower bound on convergence rate of some of the consensus algorithms is provided. Towards this end, two approaches based on the properties of stochastic matrices and the concept of random walks are used. In [70], it is shown that connectivity of a network with a fixed number of links can be significantly increased by selecting the interagent information flow links properly. This in turn can result in an ultra fast consensus-seeking procedure. The idea is best applicable to small-world networks where any two nodes can be connected using a few links, though the total size of the network can be large.

In some work, the communication delay is considered in modelling of a network of point-mass agents. For example, one can refer to [79] in which directed and undirected networks with fixed and switching topologies are considered. It is assumed that the delayed information from other agents is compared with the delayed value of the agent's own dynamics at each time step. On the other hand, in [65] the delayed information of the neighbors are compared with the current value of the agents' state. In this work, uniform time-delays in communication links are analyzed for consensus algorithms. In [52], the agreement protocol is analyzed when there are nonuniform time-delays in the links among the agents. Linear protocols are used for fixed networks with and without communication time-delays and communication channels that have filtering effects. Similarly, nonlinear protocols are applied to dynamic networks to achieve consensus. In all the above cases, the effects of time-delay are analyzed for the agreement protocol only and the analysis is performed in the frequency domain. In [68], the authors have considered time-varying time-delays in communication links and presented conditions for consensus achievement in a network of nonlinear, locally passive systems. Considering time-delays in other coordination problems such as formation control and target tracking is still an open area of research.

The problem of team cooperation and specifically consensus seeking with switching topology, has received a lot of attention in recent years and has been discussed in the literature from different perspectives [129–135]. The work performed in [135] can be considered as one of the pioneer work in which algorithms for distributed computation in a network with a time-varying network structure are analyzed. Specifically, in [136] for a discrete-time model of processors and a given number of tasks, convergence of a consensus algorithm in a time-varying structure is discussed, given that some restrictions are imposed on the frequency of availability of the interagent communication links. One of the underlying assumptions in many of the related work on switching networks is that the graph describing the information exchange structure is a balanced graph. The authors in [129] considered balanced information graphs and shown the stability under switching time-delayed communication links. The analysis is performed by introducing a Lyapunov functional and then showing the feasibility of a set of linear matrix inequalities (LMIs). In [141, 143], switching control laws are designed for a network of multi-agent systems with undirected and connected underlying graphs whereas in [49], consensus in a directed, jointly connected and balanced network is discussed. The necessary conditions for achieving consensus are discussed in [138]. The concept of pre-leader-follower is introduced as a new approach to achieve consensus in a network of discrete-time systems. The basic properties of stochastic matrices are used to guarantee consensus achievement with switching topology and time-delayed communication links.

In [89], higher order consensus algorithms are discussed. The author's approach to handle the switching network structure with a spanning tree is to determine an appropriate dwell time with its own provided definition. It is shown that the final consensus value depends on the information exchange structure as well as on the controller weights. In [66], analysis is performed for a time-varying network of multi-agent systems with discrete-time models. In this work, milder assumptions on connectivity of the agents over time are imposed when compared to [46], and necessary and sufficient conditions for consensus achieving are presented. The works in [130–132] are extensions of the approach in [46] for second-order dynamics of fixed and dynamic topologies in undirected and connected graphs. In [117], the authors have used a similar approach to the one in [130–132] to analyze consensus achievement in a team with fixed and switching topologies. They divided the control law into multiple parts and used non-smooth analysis framework to address the problem. In [79], consensus achievement for a connected graph subject to certain switching in the network structure is addressed. The underlying assumption is that the graph under consideration is a balanced graph.

In [49, 93], consensus in directed, jointly connected, and balanced network is discussed. The authors in [93] have considered information consensus in multi-agent networks with dynamically changing interaction topologies in the presence of limited information. They have shown that consensus can be achieved asymptotically under these conditions if the union of the directed interaction graphs has a spanning tree frequently enough for agents with discrete- and continuous-time dynamics. This condition is weaker than the assumption of connectedness that is made in [79] and [46] and implies that one half of the information exchange links that are required in [46] can be removed without affecting the convergence result. However, the final achieved equilibrium points will depend on the property of the directed graph, e.g., its connectedness. This work is an extension of [46] to digraphs case with more flexible weight selections in the information update schemes. Some simulation examples of this work are presented in [90].

One of the recent research topics in consensus seeking is analysis of behavior of the consensus algorithms in presence of measurement noise or even design of consensus algorithms which can compensate for the lack of measurements or inaccuracy of sensors. In [137], the performance of first- and second-order consensus algorithms is discussed in presence of measurement noise. A relationship between the measurement error and the consensus error is derived. In [60], the measurement noise is considered for the leader-follower structure. Having used the stochastic analysis and assuming time-varying weights, the authors could guarantee a mean square consensus achievement in presence of measurement noise. Similar idea of using time-varying weights is used in [56] to guarantee consensus on a Gaussian random variable. The authors in [11] have linked the consensus problem into a multi-inventory system control problem, where bounded disturbances affect the first-order agents' dynamics.

One of the approaches for solving the consensus problem that is pursued in this book is based on the decentralized optimal control theory. Next, we provide a brief literature review on the optimal control approaches to consensus-seeking problem. Among the first work where optimal control was discussed in a team of agents one can refer to [39]. There, the linear quadratic regulator (LQR) problem was solved by *using* a team of decision makers and not *in* a team of decision makers. In other words, each decision maker is responsible for design of an optimal control at one (or some) time instant where the other decision makers should decide what the best (optimal) actions for the next time instants are to minimize a common cost function. Therefore, although the problem is dynamic in the sense that at the outset an optimal controller is designed to minimize a cost function with a given dynamical constraint, none of the decision makers have an individual dynamics. This implies that each decision maker can be interpreted as the state of a discrete-time system at one time only and not as an independent dynamical system.

In contrast, in the optimal control approach to the consensus problem that is introduced in this book, each decision maker has its own continuous-time dynamics and its own cost function which is coupled to the state and control (action) of the other members. Therefore, each decision maker should decide on a *set* of actions and not only for a *single* one-time action. In other words, in this book the purpose is that N decision makers design N control actions for *all the time* period whereas in [39], N decision makers design *one* control action each for only one time instant. Moreover, decentralization in the context of [39] implies that a single control action (decision) should be designed by the involvement of several decision makers, whereas in the framework presented in this book the decentralization refers to design of several controllers where the objective of the controllers is then coupled to each other.

In more recent literature, optimal control approach to the team cooperation problem is considered in [29, 35, 36, 44, 82] for formation keeping and in [10, 24, 25, 42] for consensus seeking. The approach in [44] is based on the individual agent cost optimization for achieving team goals under the assumption that the states of the other team members are constant. The concepts of Nash equilibrium, penalty function, as well as Pareto optimality are used for design of optimal controllers. In [35], for a formation-keeping problem, the effects of the extent of information on the value of the cost are investigated. The authors have shown that a centralized architecture will result in the lowest cost whereas the decentralized solution will increase the cost value. In order to solve an optimal consensus problem, the authors in [10] have assumed an individual agent cost for each team member. In evaluating the minimum value of each individual cost, the states of the other agents are assumed to be constant. For a switching network structure the dwell time that provides stability of the network subject to the switching structure is obtained. In [42], an H_2 optimal semi-stable methodology for stabilization of linear discrete-time systems is proposed. The authors then proposed a consensus algorithm and have shown that this protocol is a semi-stable controller that can solve the consensus-seeking problem. The authors in [25] have shown that a specific type of graphs, namely the de Bruijn's graph, is optimal for consensus-seeking problem and with respect to a given cost function.

In all the above referenced work the optimal control problem is developed based on the notion of an individual cost function for team members. However, a single team cost function formulation has been proposed in only a few work [29, 82]. In [29], optimal control strategy is applied for formation keeping where a single team cost function is utilized. The authors in [82] assumed a distributed optimization technique for the formation control in a leader-follower structure. The design is based on dual decomposition of local and global constraints. However, in this approach, the velocity and position commands are assumed to be available to the entire team. In [24], the dynamics of the entire network are decomposed into two components, namely one in the consensus space and the other in its orthogonal subspace. A set of LMIs are then used to guarantee the stability and consensus achievement by using an H_2 design strategy.

1.4 General Statement of the Problem

In [95], coordination in a network of unmanned vehicles is defined as a problem which addresses one or some of the following issues: a cooperative motion such as formation control problem, network agreement, consensus seeking, flocking, or swarming behavior; or high-level cooperative control problems such as task assignment, timing, and path planning. Furthermore, the reader is referred to [146] for more information on various directions of research on coordination in networked control systems. In general, any network design problem can be formulated according to four main subproblems, namely:

1. Control design: Specifically issues related to stabilization, controllability, observability, and robustness
2. Network structure design: Specifically issues related to minimization of communication required to achieve a predefined goal by determining conditions on the structure or on different strategies for information exchanges
3. Decentralized/distributed estimation: Specifically issues related to stability and convergence analysis
4. Low-level communication problems: Specifically issues related to congestion, routing, and resource allocation

The main challenges in addressing and solving these problems are:

- Real-time sensor processing and decision making: Specifically as they relate to real-time implementation requirements
- Communication constraints: Specifically as they relate to delayed and missed information due to bandwidth limitations and time-varying communication structures

- Computational constraints and requirements
- Design of a distributed collaborative team consisting of agents with more complex dynamics than point-mass models and incomplete access to information of the entire team
- Distributed estimation protocols in the network
- Analysis of team cooperative behavior specifically as it relates to presence of faults and anomalies in the team members

Despite the fact that a large body of work has already appeared in the literature to accomplish the coordination requirements and to address the above problems, there are still unsolved problems in this area. Some of these problems are listed below:

1. Complete information is not provided to all the agents. In a leader-follower structure and in certain circumstances, the leader command is not known to the followers. Moreover, in any given structure it is possible that an agent is only aware of its neighbors' states. The team goals such as cohesion and connectivity should still be accomplished even if the full information set of the team is not available to the entire team due to the presence of uncertainties and partial communication failures.
2. Most of the results that have appeared in the literature deal with a point-mass model of multi-agent systems; see [46,73–75]. Extension of these results to more complex and realistic agents' dynamics is an active and open area of research.
3. Robustness issues and considerations for various network structures need to be investigated. Consideration of scenarios such as failure of the leader in a leader-follower structure, recovery of the team in a failure mode, and fault diagnosis and recovery in the formation-keeping or consensus-seeking problems should be addressed.
4. Dynamic network topologies with time-varying structures and changing number of nodes in the network should indeed be considered.
5. The behavior of the resulting team following the presence of agents' faults and malfunctions should be predicted and analyzed.

In view of the above, the main objective and purpose of this book is to design a team of multi-agent systems that can accomplish and fulfill several goals and requirements under constraints. The team should be able to perform various maneuvers corresponding to several scenarios that are required in different applications. The main focus is on design of consensus protocols for a distributed network of multi-agent systems such that the team operates and works cooperatively for achieving a cohesive motion. There are certain constraints on the availability of information as well as on the availability of communication links during the mission. Therefore, time-varying network topology is addressed as well. In order to make the problem more practical, it is assumed that anomalies could occur in the team members which will lead to deteriorations in the performance of the overall team.

In this book a new formulation of the problem is introduced to address the above important issues. The problem of designing cooperating multi-agent systems within a distributed information network, known as the team collaboration in control engineering and economics, is of considerable importance and plays a key role in understanding complex systems. Specifically, the main focus of this book is to introduce a unified methodology for solving the team cooperation problem in a framework which is sufficiently broad to allow addressing and considering it subject to various types of constraints and limitations. The main challenging aspect of the problem is that the autonomy of the team should be satisfied in spite of the fact that no supervisor or command manager is present and moreover even the overall team information may not be available to all the agents. Specifically, the relative requirements (such as in regulation of relative distances or velocities) that are imposed on the agents do indeed introduce challenging issues for research and development of corresponding solutions. In other words, the nature of the problem is such that the decision makers should be designed based on a (semi-)decentralized architecture, while on the other hand due to the autonomy and cooperative nature of the multi-agent systems, some of the agents may not have access to the exact set point or tracking path for their decision making. The proposed team cooperative strategy accomplishes agreement or command-tracking goals by using optimal control, game theory, and LMI methodologies through the introduction of interaction terms to overcome the relative specifications and dependencies of individual agent goals on other agents' outputs or states.

1.5 Outline of the Book

This book is organized as follows: In Chap. 2, the required background for the following chapters is provided. Chapter 3 is devoted to the design of semi-decentralized optimal control strategies based on the solution of the Hamilton–Jacobi–Bellman (HJB) equations. In Chap. 4, two realistic and critical scenarios are discussed corresponding to the strategies that are proposed in Chap. 3. In the first scenario, a failure analysis is performed when some of the agents in the team are subject to actuator faults. The second scenario deals with the switching control design and stability analysis for a dynamic network of mobile robots where both the network structure and the leader assignment are possibly switching. In Chap. 5, consensus-seeking problem is solved by using a game-theoretic framework as well as an optimal control approach by utilizing the LMI formulation. Corresponding to the optimal control solution, the notions of state decomposition and LMI formulation are used. Finally, the book is concluded in Chap. 6 where directions for future work are provided.

Chapter 2
Background

In this chapter, all the background material that is required for the remainder of this book are provided. These topics cover formal definition of a network of multi-agent systems as well as the dynamical model of an agent, communication structures, and the formal statement of the consensus-seeking problem in a network of multi-agent systems. Other sections provide background on mathematical frameworks that are used in this book. Examples are introduction to the game theory framework, the optimal control problem formulation, as well as linear matrix inequalities (LMIs).

2.1 A Network of Multi-agent Systems

In general, an agent refers to a dynamical system. However, in the context of this book, the term "agent" is interchangeable with "vehicle," where a vehicle may be a mobile robot or any other vehicle whose dynamics can be described by linear time-invariant models.

Assume a set of agents $\Omega = \{i = 1,\ldots,N\}$, where N is the number of agents. Each member of the team, which is denoted by i, is placed at a vertex of the network information graph. In general, the dynamical representation of each agent is governed by

$$\dot{X}^i = A^i X^i + B^i u^i,\ Y^i = c^i X^i,\ i = 1,\ldots,N, \tag{2.1}$$

where $X^i \in R^q$, $u^i \in R^m$, and $Y^i \in R^n$ are the state, input, and output vectors of the agent i. The corresponding vectors of the entire team are designated by X, U, and Y, respectively, which are the concatenation of all the state, input, and output vectors of the agents and are given by

$$\begin{aligned} X_{Nq\times 1} &= [(X^1)^T \ \ldots \ (X^N)^T]^T,\ U_{Nm\times 1} = [(u^1)^T \ \ldots \ (u^N)^T]^T, \\ Y_{Nn\times 1} &= [(Y^1)^T \ \ldots \ (Y^N)^T]^T. \end{aligned} \tag{2.2}$$

E. Semsar-Kazerooni and K. Khorasani, *Team Cooperation in a Network of Multi-Vehicle Unmanned Systems: Synthesis of Consensus Algorithms*,
DOI 10.1007/978-1-4614-5073-3_2, © Springer Science+Business Media New York 2013

2.2 Information Structure and Neighboring Sets

In order to ensure cooperation and coordination among the members of a team, each member has to be aware of the status of other members (i.e., the output or the state vector), and therefore the members have to communicate with one another. For a given agent i in a network with an undirected underlying graph, the set of agents from which it can receive information is called a neighboring set N^i, i.e.,

$$\forall\, i = 1,\ldots,N, \;\; N^i = \{j = 1,\ldots,N \mid (i,j) \in E\}, \tag{2.3}$$

where E is the edge set that corresponds to the underlying graph of the network. In other words, i and j either have a communication link to transfer their status (output vector) to each other or have another way to obtain information such as measurements of other agents' states. Furthermore, all the team members have two-way links with agents that are connected to them. Based on this formulation the number of neighbors of the agent i is $|N^i|$ (the cardinality of the set N^i).

2.2.1 *Leaderless Structure*

Figure 2.1a shows the underlying graph of a leaderless information exchange structure. In this figure, the circles stand for the agents and the arrows stand for the communication links. In the leaderless (LL) structure, each agent communicates with a group of agents, i.e., its neighbors, without following any specific agent as its leader. Subject to utilizing an appropriate control strategy, eventually the agents' output, e.g., their velocity, converge to a common value which is decided upon by

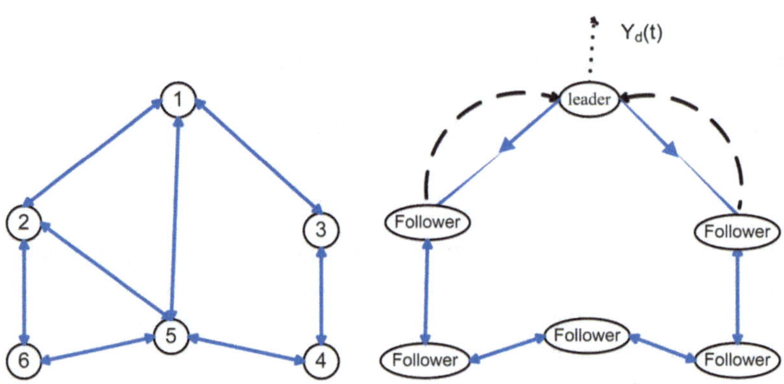

Fig. 2.1 Information structure in **(a)** a leaderless structure and **(b)** a modified leader–follower structure

the team members, i.e., $\forall i, j\ Y^i \rightarrow Y^j$. This state is referred to as the consensus state. Hence, although no desired command is a priori given to the agents, they can reach a consensus by using the existing interconnections.

2.2.2 Modified Leader-Follower Structure

Coordination problems may be treated in a different architecture in which the desired value to which the agents should converge is provided to the team via a supervisor or commander, which can be either a leader as in the leader-follower structure [118, 133, 134] or a coordination vector as in the virtual structure configuration [92]. In the standard leader-follower configuration, the leader can affect the followers whenever it is in their neighboring set, but there is *no* feedback from the followers to the leader. In [12, 13, 92], the idea of adding a feedback from the virtual structure to the coordination unit is introduced. Other than leader-follower and virtual structure there can be an alternative architecture in which the command information is available to part of the team, e.g., to the leader, and the rest of the team should receive this information via links that exist among the neighbors. The work in [46] can be considered as an example of this structure for a team of multi-agent systems with single integrator dynamical model where the leader can affect the followers but there is no feedback from the followers to the leader. Using the nearest neighbor rule, "each agent's heading is updated using the average of its own heading plus the headings of its neighbors [46]."

In this book, a modified leader-follower (MLF) structure is introduced and utilized. The difference between the MLF architecture that is introduced here and the above mentioned architectures is that here only the leader is aware of the command, and the rest of the team is connected to each other or to the leader through a predefined topology. The command can be a set point reference or a time-varying signal specified for the output or a trajectory to be followed by the agents. Here, the leader control input *can* also be affected by the followers through a corrective feedback from the followers.

To be more specific, assume an information exchange structure as shown in Fig. 2.1b. An external command is provided to one of the agents that is designated as the leader, and the objective is to make the agents' output, e.g., velocity, converge to the external command, i.e., $\forall i,\ Y^i \rightarrow Y^d$. Other agents should follow the leader by communicating through their links with each other or with the leader. Furthermore, the leader has two-way links with the followers that are connected to it, i.e., it receives feedback from *some* of the followers. This implies that although the leader may follow a given trajectory without any feedback from the followers, it receives the status of the followers, which may contribute to improving the robustness of the team cohesion, as shown and developed formally in Chap. 4.

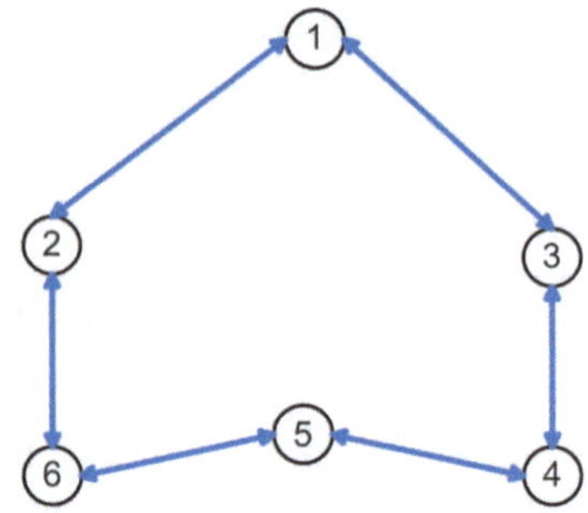

Fig. 2.2 Information structure in a ring topology

2.2.3 Ring Topology

In this topology, as can be seen in Fig. 2.2, each agent is connected to its two adjacent neighbors, i.e.,

$$\forall i = 2,\ldots,N-1,\ N^i = \{i-1, i+1\},\ N^N = \{N-1, 1\},\ N^1 = \{2, N\}, \qquad (2.4)$$

and the corresponding Laplacian matrix, as will be defined in Sect. 2.5, is given by

$$L = \begin{bmatrix} 2 & -1 & 0 & \ldots & 0 & -1 \\ -1 & 2 & -1 & 0 & \ldots & 0 \\ \vdots & \vdots & \vdots & \vdots & \vdots & \\ -1 & 0 & \ldots & 0 & -1 & 2 \end{bmatrix}. \qquad (2.5)$$

2.3 Model of Interactions Among the Team Members

In decentralized control, the multilevel control concept is used as a means to compensate for the coupling dynamics in an interconnection of several subsystems. In the simplest case, the control u^i is partitioned into two parts, namely local and global components, i.e., $u^i = u^i_g + u^i_l$. The approach pursued here is clearly different from what is used in standard decentralized control, even though the general idea seems to be similar. Specifically, first an interconnection (interaction) term is introduced in the agent's dynamics in order to represent the connectivity of the members as required by the team objective. The interaction terms play the role of a global controller in the decentralized approach, so that they can eliminate the effects of relative dependency in design of the individual controllers.

To formalize this concept, assume that the dynamical model of each agent is given by (2.1). This model defines the dynamics of an isolated agent of the team; however, team agents do indeed have certain interactions among their neighboring agents through existing information flows. This interaction may be considered for each agent through its input channel and the available information from outputs of other agents in its neighboring set. These concepts may be represented by the following dynamical model:

$$\dot{X}^i = A^i X^i + B^i u_l^i(X^i) + B^i u_g^i(Y^j),\ Y^i = c^i X^i. \tag{2.6}$$

In other words, the actual control input $u^i(X^i, Y^j), i = 1, \ldots, N$ is decomposed into two components

$$u^i(X^i, Y^j) = u_l^i(X^i) + u_g^i(Y^j),\quad j \in N^i,$$

in which $u_g^i(Y^j)$ defines the dependence of the control input of agent i on its neighbors' information explicitly and $u_l^i(X^i)$ defines the dependence of the control input of the agent i on its local information. The controllers u_l^i have the same role as local controllers in the decentralized approach. For simplicity, we assume that $u_g^i(Y^j) = \sum_{j \in N^i} \mathscr{F}^{ij} Y^j,\ i = 1, \ldots, N$, where $\mathscr{F}^{ij}$ is the interaction matrix to ensure compatibility of the agent's input and output channels dimensions. The above formulation illustrates why the method that will be proposed in the following chapters will be called semi-decentralized. Whereas u_l^i is only a function of X^i, u_g^i incorporates the effects of neighbors' information (output) into the control law so that the control input is not fully decentralized in the conventional sense.

By incorporating the interaction terms in the agent's dynamics, the representation model of the entire network can be written as follows:

$$\dot{X} = AX + BU,\quad Y = CX, \tag{2.7}$$

in which X and Y are the entire team state and output vectors as defined previously. Vector U is defined as in (2.2) by replacing u^i with u_l^i for all i. Matrices A, B, and C are defined as follows:

$$A = \begin{bmatrix} A^1,\ 0, \ldots, B^1 \mathscr{F}^{1j} c^j, \ldots, 0 \\ \vdots \\ 0, \ldots, B^N \mathscr{F}^{Nj} c^j, \ldots, 0, A^N \end{bmatrix},\ B = \mathrm{Diag}\{B^1,\ \ldots,\ B^N\},$$

$$C = \mathrm{Diag}\{c^1, \ldots, c^N\}. \tag{2.8}$$

The terms $B^i \mathscr{F}^{ij} C^j$ represent the interactions that exist among the agents.

Remark 2.1. The phrase "interaction terms" is also commonly used in decentralized control of large-scale systems and refers to part of the dynamics which describes the existing coupling terms among the subsystems. However, the interaction terms in the context of this book refers to the *externally added* couplings that are added in the input channels and as part of the control law. The proposed interactions guarantee that the team goals and specifications are satisfied as opposed to compensating for the dynamical couplings among the agents.

2.4 Dynamical Model of an Agent

In this section, we will discuss and introduce two linear dynamical representations for the agents. These models will be used in our future discussions. It is worth noting that many of the mobile vehicles have nonlinear dynamical equations due to their existing non-holonomic constraints [4, 20]. However, it will be shown in the following sections that a common nonlinear model of a mobile robot can be transformed into a linear one by using a nonlinear feedback of the states. Therefore, the assumption of linear dynamical model for mobile robots is reasonable even if the vehicles have non-holonomic constraints.

2.4.1 *Mobile Robot Dynamical Model: Double Integrator Dynamical Model*

In this section, the agents are mainly assumed to be mobile robots. There are different models for mobile robots in the literature. In [4], a seventh-order model is considered for a mobile robot with non-slipping and pure-rolling motion. The robots are considered to have two steering wheels and a free one with non-holonomic constraints. Based on some simplifications that are discussed in [4], and due to the planar motion, these equations can be simplified into the following fifth-order model:

$$\begin{aligned}
\dot{x}^i &= \eta_1^i \sin\theta^i, \\
\dot{y}^i &= \eta_1^i \cos\theta^i, \\
\dot{\theta}^i &= \eta_2^i, \\
\dot{\eta}_1^i &= a_1^i = \frac{1}{M^i}\mathbb{F}^i, \\
\dot{\eta}_2^i &= a_2^i = \frac{1}{\mathbb{J}^i}\tau^i,
\end{aligned} \tag{2.9}$$

in which x^i and y^i are the position components, η_1^i and η_2^i are the linear and the angular velocities, θ^i is the heading angle, a_1^i and a_2^i are the linear and angular accelerations, $\mathbb{F}^i$ and τ^i are the force and torque inputs, and M^i and $\mathbb{J}^i$ are the mass and moment of inertia of each robot, respectively.

The above model is in general a nonlinear model; however, as the focus of this book is on linear models, in the following we will show how to linearize this model. For certain definitions of the output this model can be transformed exactly into a linear model. Three cases are assumed for linearizing the model based on different output definitions:

Case I: Output of each agent is one of the three last states or a combination of them. For example, assume that $Y^i = [\theta^i \;\; \eta_1^i]^T$. Based on this output vector definition the first two equations may be eliminated as they do not have any effect on the input-output channel. Hence, a reduced order model may be considered which is given by

$$\begin{aligned} \dot{\theta}^i &= \eta_2^i, \\ \dot{\eta_1}^i &= a_1^i = \frac{1}{M^i}\mathbb{F}^i, \\ \dot{\eta_2}^i &= a_2^i = \frac{1}{J^i}\tau^i . \end{aligned} \tag{2.10}$$

The new state vector may now be considered as $X^i = [\theta^i, \; \eta_1^i, \; \eta_2^i]^T$.

Case II.a: Output of each agent consists of at least one of the first two states, namely

$$Y^i = [x^i \;\; y^i]^T . \tag{2.11}$$

In this case the first two equations cannot be separated anymore. Hence, one has to invoke the feedback linearization procedure. Based on the notion of the relative degree that is presented in [48], the relative degree of the system in (2.9) with respect to the output in (2.11) is $\{2,2\}$. However, the system is not feedback linearizable. Hence, we need to modify the input-output definition. One common solution is to consider a dynamic input definition. Towards this end, define $z^i = [\dot{a}_1^i \;\; a_2^i]^T$ as the new input vector. Hence, (2.9) can be written as

$$\begin{aligned} \dot{x}^i &= \eta_1^i \sin\theta^i, \\ \dot{y}^i &= \eta_1^i \cos\theta^i, \\ \dot{\theta}^i &= \eta_2^i, \\ \dot{\eta}_1^i &= a_1^i = \frac{1}{M^i}\mathbb{F}^i, \\ \dot{a}_1^i &= z_1^i, \\ \dot{\eta}_2^i &= z_2^i = \frac{1}{J^i}\tau^i . \end{aligned} \tag{2.12}$$

The above system is now feedback linearizable with the relative degree $\{3,3\}$. By utilizing the state transformation

$$X = \begin{bmatrix} x^i \\ y^i \\ \eta_1^i \sin\theta^i \\ \eta_1^i \cos\theta^i \\ \eta_1^i \eta_2^i \cos\theta^i + a_1^i \sin\theta^i \\ -\eta_1^i \eta_2^i \sin\theta^i + a_1^i \cos\theta^i \end{bmatrix}, \tag{2.13}$$

the system can be transformed into the linear model (2.1) with the following set of parameters:

$$A^i = \begin{bmatrix} 0_{2\times2} & I_{2\times2} & 0_{2\times2} \\ 0_{2\times2} & 0_{2\times2} & I_{2\times2} \\ 0_{2\times2} & 0_{2\times2} & 0_{2\times2} \end{bmatrix}, B^i = \begin{bmatrix} 0_{2\times2} \\ 0_{2\times2} \\ I_{2\times2} \end{bmatrix}, c^i = \begin{bmatrix} I_{2\times2} & 0_{2\times2} & 0_{2\times2} \end{bmatrix}, \tag{2.14}$$

and

$$u^i = \begin{bmatrix} \sin\theta^i & \eta_1^i \cos\theta^i \\ \cos\theta^i & -\eta_1^i \sin\theta^i \end{bmatrix} z^i + \eta_2^i \begin{bmatrix} 2a_1^i \cos\theta^i - \eta_1^i \eta_2^i \sin\theta^i \\ -2a_1^i \sin\theta^i - \eta_1^i \eta_2^i \cos\theta^i \end{bmatrix}. \tag{2.15}$$

The coefficient matrix of z^i is nonsingular for $\eta_1^i \neq 0$, and therefore the control law for u^i can be transformed to a value for z^i and then to the v^i (defined below). The model presented in (2.1) with parameters as in (2.14) is a chain of integrators. It is simple to verify that whereas the first two states in the new coordinates are the position states, the third and the fourth states are linear velocity components and the last two states are the acceleration components. If we denote the position, the velocity, and the acceleration vectors of each agent by r^i, v^i, and a^i, respectively, it can be shown that

$$\begin{aligned} \dot{r}^i &= v^i, \\ \dot{v}^i &= a^i, \\ \dot{a}^i &= u^i. \end{aligned} \tag{2.16}$$

Case II.b: The definition of the output is similar to the previous case, but here the first four states are assumed as the state vector and the input vector is $z^i = [a_1^i \;\; \eta_2^i]^T$. This can be justified as the dynamics of η_2^i does not have any effect on the input-output channel. In this case, the relative degree of the system is $\{2,2\}$. By introducing the following state transformation

$$X = \begin{bmatrix} x^i \\ y^i \\ \eta_1^i \sin\theta^i \\ \eta_1^i \cos\theta^i \end{bmatrix}, \tag{2.17}$$

the system can be transformed into the linear model (2.1) with the following parameters:

$$A^i = \begin{bmatrix} 0_{2\times2} & I_{2\times2} \\ 0_{2\times2} & 0_{2\times2} \end{bmatrix}, B^i = \begin{bmatrix} 0_{2\times2} \\ I_{2\times2} \end{bmatrix}, c^i = \begin{bmatrix} I_{2\times2} & 0_{2\times2} \end{bmatrix}, \tag{2.18}$$

and $u^i = \begin{bmatrix} \sin\theta^i & \eta_1^i \cos\theta^i \\ \cos\theta^i & -\eta_1^i \sin\theta^i \end{bmatrix} z^i$. The coefficient matrix of z^i is nonsingular for $\eta_1^i \neq 0$, and therefore the control law for u^i can be transformed to a value for z^i and then a_1^i and η_2^i can be found. The model that is presented in (2.1) with parameters as in (2.18) is again a chain of integrators. It is simple to verify that while the first two states in the new coordinates are the position states, the last two states are velocity components. If one denotes the position and the linear velocity of each agent by r^i and v^i, respectively, then

$$\begin{aligned} \dot{r}^i &= v^i, \\ \dot{v}^i &= u^i,\ i = 1,\dots,N. \end{aligned} \tag{2.19}$$

To summarize the above discussion, we assume that the vehicles are mobile robots that are represented by double integrator dynamical models as in (2.19) and that $Y^i = v^i, i = 1,\dots,N$. Using the formulation proposed in Sect. 2.3, the dynamical model of each agent with interaction terms is as follows:

$$\begin{aligned} \dot{r}^i &= v^i, \\ \dot{v}^i &= u_l^i + u_g^i,\ u_g^i = \sum_{j \in N^i} \mathscr{F}^{ij} v^j, \\ Y^i &= v^i, i = 1,\dots,N. \end{aligned} \tag{2.20}$$

2.4.2 Linear Dynamical Model

As discussed in the previous section, in general the dynamical equations of most ground vehicles can be transformed into a generalized form of a set of integrators (canonical form). Given the focus of this work on application of teams of mobile robots, our assumption on the agent dynamical equations is an admissible one. In this book, we assume that the vehicles' dynamics consist of position and velocity, i.e., $X^i = [X_1^i\ X_2^i]^T = [(r^i)^T\ (v^i)^T]^T$. Therefore, in the presence of interaction terms, the following agents' dynamical equations are considered:

$$\begin{aligned} \dot{r}^i &= \bar{A}^i v^i, \\ \dot{v}^i &= A^i v^i + B^i(u_l^i + u_g^i),\ u_g^i = \sum_{j \in N^i} \mathscr{F}^{ij} Y^j, \\ Y^i &= v^i \in R^n,\ i = 1,\dots,N\,, \end{aligned} \tag{2.21}$$

where B^i is a nonsingular matrix, A^i and $\bar{A}^i$ are $n \times n$ matrices, and r^i, v^i denote the ith agent position and velocity vectors, respectively.

Remark 2.2. The dynamical model of each agent consists of position and velocity states. However, since the main objective in this work is to have a common output, namely velocity, for analysis we will only consider the velocity dynamics to describe the dynamical behavior of each agent. It should be noted that for the purpose of simulation studies naturally the position dynamics of each agent is also included.

2.5 Terminologies and Definitions

In the following we define relevant terminologies that are commonly used in the literature on graph theory:

- *Laplacian Matrix:* [30] This matrix is used to describe the graph $\mathscr{G}$ associated with the information exchanges in a network of multi-agent systems and is defined as $L = [l_{ij}]_{N\times N}$, where

$$l_{ij} = \begin{cases} d(i) & i = j \\ -1 & (i,j) \in E \text{ and } i \neq j \\ 0 & \text{otherwise,} \end{cases} \tag{2.22}$$

  where E is the edge set of the graph $\mathscr{G}$, $d(i)$ is equal to the cardinality of the set N^i [76], $|N^i|$ is called the degree of vertex i. For an undirected graph, the degree of a vertex is the number of edges incident to that vertex (total number of links connected to that vertex). For directed graphs, instead of the degree, either the in-degree or the out-degree might be used (the total number of links that are entering or leaving a node).
- *Normalized Laplacian Matrix:* The normalized Laplacian matrix $\hat{L}$ is defined similar to the Laplacian matrix of a graph, where $\hat{L} = [\hat{l}_{ij}]_{N\times N}$ and

$$\hat{l}_{ij} = \begin{cases} 1 & i = j \\ -1/d(i) & (i,j) \in E \text{ and } i \neq j \\ 0 & \text{otherwise.} \end{cases} \tag{2.23}$$

- *Adjacency Matrix:* [34] The adjacency matrix of a graph, denoted by $\mathbf{A}$, is a square matrix of size N, defined by $\mathbf{A}(i,j) = 1$ if $(i,j) \in E$ and $i \neq j$, and is zero otherwise.
- *Normalized Adjacency Matrix:* The normalized adjacency matrix of a graph, denoted by $\hat{\mathbf{A}}$, is a square matrix of size N, defined by $\hat{\mathbf{A}}(i,j) = \frac{1}{d(i)}$ if $(i,j) \in E$ and $i \neq j$, and is zero otherwise.
- *Connected Graph:* [67] An undirected graph consisting of a vertex set, V, and an edge set, E, is connected if there is a path between any two vertices and the path lies in the edge set.
- *Balanced Graph:* If the Laplacian matrix of a graph, L, has the property that $\mathbf{1}^T L = 0$, then the graph is called a balanced graph. For balanced connected

graphs one has the property that $L+L^T$ can be considered as the Laplacian matrix of an undirected and connected graph [79].

- *Tree:* [34] A connected graph with no cycles (acyclic) is called a tree.
- *Spanning Tree:* [34] A spanning tree of a connected undirected graph is a connected subgraph of the original graph with the same vertex set as the original graph and no cycles.
- *Forest:* [34] A forest is a disjoint union of trees.
- *Path:* [147] In a tree, every two vertices are connected by a unique path. The length of this path is the distance between the two vertices.

The followings are some other useful definitions from the linear algebra literature:

- *Spectrum of a Matrix:* [41] The set of eigenvalues of a matrix is called its spectrum.
- *Inertia of a Hermitian Matrix:* [41] is the number of positive, negative, and zero eigenvalues of a matrix.
- *Spectral Radius:* [41] The spectral radius of a matrix A, i.e., $\rho(A)$, is the largest of the absolute value of the eigenvalues (or magnitude of the complex eigenvalues) of that matrix, i.e.,

$$\rho(A) = \max_i |\lambda_i| \,, \tag{2.24}$$

where λ_i is an eigenvalue of the matrix A.

Definition 2.1 ([17] Schur Complement). Assume that matrix M is given by $M = \begin{pmatrix} A & B \\ C & D \end{pmatrix}$, where D is invertible. The Schur complement of the block D of the matrix M is $A - BD^{-1}C$. This conversion can then be used to transform bilinear (nonlinear) matrix inequalities into LMIs, i.e.,

$$\begin{pmatrix} A & B \\ B^T & D \end{pmatrix} \geq 0,\ D > 0 \Leftrightarrow A - BD^{-1}B^T \geq 0\,. \tag{2.25}$$

The following is the well-known Perron–Frobenius theorem for nonnegative matrices. This theorem will be used in Chapter 4.

Theorem 2.1 (Perron–Frobenius Theorem [41]). *Let $M \in M_n$ and suppose that M is irreducible and nonnegative. Then*

(1) $\rho(M) > 0$.
(2) $\rho(M)$ is an eigenvalue of M.
(3) There is a positive vector x such that $Mx = \rho(M)x$.
(4) $\rho(M)$ is an algebraically simple eigenvalue of M.

where M_n is the set of matrices of order n and $\rho(M)$ is the spectral radius of the matrix M [41].

Definition 2.2. In a network consisting of nodes that are denoted by i and corresponding to the team system dynamics as $\dot{x} = f(x)$, $x = [x_1^T, \ldots, x_i^T, \ldots, x_N^T]^T$, a protocol asymptotically solves the χ-consensus problem if and only if there exists an asymptotically stable equilibrium x^* satisfying $x_i^* = \chi(x(0))$ for all the nodes i [79], where $x(0)$ is the initial value of the state vector x. If $\chi(x) = \text{Ave}(x) = \frac{1}{N}\sum_{i=1}^{N} x_i$, the protocol is called average consensus. Similarly, other χ-consensus protocols such as max-consensus and min-consensus can be defined.

2.6 Types of Actuator Faults

In [15], the following three types of actuator faults are introduced that are of interest in this book, namely

1. *Lock-in-place (LIP)*: In this case, the actuator freezes at a certain value and does not respond to the subsequent commands.
2. *Float*: The float fault occurs when the actuator output floats with zero value and does not contribute to the control authority.
3. *Loss of effectiveness (LOE)*: The LOE fault is characterized by lowering the actuator gain with respect to its nominal value.

These faults may be formally represented as follows:

$$u_f^i = \begin{cases} \alpha u^i & \alpha = 1,\ \forall t \geq 0 & \text{No failure} \\ \alpha u^i & 0 < \varepsilon < \alpha < 1,\ \forall t \geq t_f & \text{LOE} \\ 0 & \alpha = 1,\ \forall t \geq t_f & \text{Float} \\ u^i(t_f) & \alpha = 1,\ \forall t \geq t_f & \text{LIP}, \end{cases} \tag{2.26}$$

where u_f^i corresponds to the actual input that is produced by the actuator, u^i is the input commanded by the controller, t_f denotes the time when a fault is injected, and α represents the effectiveness coefficient of the actuator and is defined to be $\alpha \in [\varepsilon, 1]$, $\varepsilon > 0$.

2.7 Hamilton–Jacobi–Bellman Equations

Minimization of a general nonlinear cost function, either unconstrained or subject to certain constraints, may be solved by using the Hamilton–Jacobi–Bellman (*HJB*) equations. In this book, the unconstrained minimization problem is considered. The *HJB* equations in the general form for a finite horizon problem are described below [9, 26].

Assume that the model of a dynamical system is given by

$$\dot{X}^i = f^i(t, X^i, u^i), \tag{2.27}$$

and the cost function to be minimized is given by

$$J^i = \int_0^T g^i(t,X^i,u^i)\mathrm{d}t + h^i(X^i(T)). \tag{2.28}$$

The solution to the following optimization problem

$$\min_{u^i \in \mathscr{U}^i} J^i = \int_0^T g^i(t,X^i,u^i)\mathrm{d}t + h^i(X^i(T))$$
$$\text{s.t. } \dot{X}^i = f^i(t,X^i,u^i)$$

is obtained if the following *HJB* equations have a solution

$$\begin{aligned} -\frac{\partial V^i}{\partial t}(t,X^i) &= \min_{u^i \in \mathscr{U}^i} \Lambda^i(t,X^i,u^i), \\ \Lambda^i(t,X^i,u^i) &= g^i(t,X^i,u^i) + \frac{\partial V^i}{\partial X^i}(t,X^i) f^i(t,X^i,u^i), \\ u^{i*}(t,X^i) &= \arg\min_{u^i \in \mathscr{U}^i} \{\Lambda^i(t,X^i,u^i)\},\ V^i(T,X^i) = h^i(X^i(T)) \end{aligned} \tag{2.29}$$

where V^i is the value function to be chosen such that the above partial differential equation (PDE) is satisfied and $\mathscr{U}^i$ is the set of all strategies for the player i. It is an indicator of the minimum value of the cost function (2.28) at any time t, i.e.,

$$V^i(t) = \int_t^T g^i(t,X^i,u^i)\mathrm{d}t + h^i(X^i(T))\,.$$

The term $h^i(X^i(T))$ is known as the "cost to go" and is the final value of the V^i, i.e., $V^i(T)$. An important issue to consider is the existence of a solution corresponding to the above equations and how the value function V^i is to be selected to satisfy the *PDE* given in (2.29).

2.8 LMI Formulation of the Linear Quadratic Regulator Problem

Consider the following cost function:

$$d = \textstyle\int_0^\infty \{X^T QX + U^T RU\}\mathrm{d}t, \tag{2.30}$$

where Q is a positive semi-definite (PSD) matrix, R is a positive definite (PD) matrix, and the corresponding linear dynamical system is given by

$$\dot{X} = AX + BU. \tag{2.31}$$

The problem of minimizing the cost function (2.30) subject to the dynamical constraint (2.31) is an linear quadratic regulator (*LQR*) problem. The solution to this minimization problem can be found if the following algebraic Riccati equation (ARE) has a solution for P:

$$PA + A^T P - PBR^{-1}B^T P + Q = 0, \tag{2.32}$$

where $U = -R^{-1}B^T PX$.

In [1, 17] it is shown that the LQR problem can be formulated as a minimization or a maximization problem that is constrained to a set of matrix inequalities. In other words, instead of solving the ARE (2.32), a set of matrix inequalities can be solved. Using this formulation, the controller $U = KX$ that minimizes the cost function (2.30) subject to (2.31) is achieved by solving for and determining the appropriate matrix P

$$\begin{aligned}
&\min X(0)^T PX(0) \text{ s.t.} \\
&PA + A^T P - PBR^{-1}B^T P + Q \leq 0, \\
&P \geq 0,
\end{aligned} \tag{2.33}$$

where $X(0)$ is the initial value of the state vector. Equivalently, the following formulation in terms of P and K can be written as

$$\begin{aligned}
&\min X(0)^T PX(0) \text{ s.t.} \\
&P(A+BK) + (A+BK)^T P + Q + K^T RK \leq 0, \\
&P \geq 0,
\end{aligned} \tag{2.34}$$

where $K = -R^{-1}B^T P$ yields the optimal solution. In [45] it is shown that if instead of the cost function (2.30) its expected value is considered and certain assumptions on the initial conditions of the system are imposed, the above minimization problem reduces to

$$\begin{aligned}
&\min \operatorname{trace}(P) \text{ s.t.} \\
&P(A+BK) + (A+BK)^T P + Q + K^T RK \leq 0, \\
&P \geq 0,
\end{aligned} \tag{2.35}$$

which can be transformed into an *LMI* optimization problem by introducing new variables $\bar{X} = P^{-1}$ and $\bar{Y} = KP^{-1}$ [45].

The following is another formulation that can be used for this purpose using a semi-definite programming framework [1], namely

$$\begin{aligned}
&\max \operatorname{trace}(P) \text{ s.t.} \\
&PA + A^T P - PBR^{-1}B^T P + Q \geq 0, \\
&P \geq 0\,,
\end{aligned} \tag{2.36}$$

where the optimal control law is then selected as $U = -R^{-1}B^TPX$. By using the Schur complement decomposition and given that $R > 0$, this formulation can be translated into an *LMI* maximization problem.

2.9 Cooperative Game Theory

In this section, we will provide a general description of the cooperative game theory and in Chap. 5, we will modify the formulation that is introduced here to make it compatible with our specific problem, i.e., the consensus-seeking problem.

Assume a team of N players with the following dynamical model:

$$\dot{x} = Ax + \sum_{i=1}^{N} B^i u^i, \tag{2.37}$$

where x is the state vector of the entire team, u^i is the individual control input, and the matrix A has an arbitrary structure. Each player or agent wants to optimize its own cost

$$J^i = \int_0^T (x^T Q^i x + (u^i)^T R^i u^i)\mathrm{d}t, \tag{2.38}$$

in which Q^i and R^i are symmetric matrices and R^i is a *PD* matrix.

If the players/agents decide to minimize their cost in noncooperative manner, a strategy (control input u^i) chosen by the ith player can increase the cost of the other players through the dynamics of the system that couples different players together. However, if the players decide to cooperate, the individual costs may be minimized. In other words, if each agent is aware of the others' decisions the agents can reduce their team cost by selecting a suitable cooperative strategy. Hence, in a cooperative strategy depending on which agent requiring extra resources the resulting minima can be different. The cooperation ensures that the total cost of the team is lower than any other noncooperative optimal solution obtained.

In a cooperative approach, it is intuitively assumed that if a set of strategies result in a lower cost for all the team members, all the players will switch to that set. Hence, by excluding this scenario, the set of desired solutions consists of those strategies that if the team changes its strategy to another one at least one of the players ends up with a higher cost. In other words, there is no alternative strategy that improves all the members' cost simultaneously. This property is formally characterized and specified in [26] as follows.

Definition 2.3 (Pareto-Efficient Strategies [26]). A set of strategies $U^* = [u^{1*}, \ldots, u^{N*}]$ are Pareto efficient if the set of inequalities

$$J^i(U) \leqslant J^i(U^*), \quad i = 1, \ldots, N \ ,$$

with at least one strict inequality does not have a solution for U. The point $J^* = [J^1(U^*), \ldots, J^N(U^*)]$ is called a Pareto solution. This solution is never fully dominated by any other solution.

Now consider the following optimization problem and assume that $Q^i \geq 0$, specifically

$$\min_{u^i \in \mathscr{U}^i} J^i = \int_0^T (x^T Q^i x + (u^i)^T R^i u^i)\mathrm{d}t$$

$$\text{s.t. } \dot{x} = Ax + \sum_{i=1}^{N} B^i u^i,$$

where $\mathscr{U}^i$ is the set of all strategies for the player/agent i. The above is a convex optimization problem. It can be shown that the following set of strategies results in a set of Pareto-efficient solutions for this problem. In other words, the solution to the following minimization problem cannot be dominated by any other solution

$$U^*(\alpha) = \arg\min_{U \in \mathscr{U}} \sum_{i=1}^{N} \alpha^i J^i(U) \,, \tag{2.39}$$

where $\alpha \in \mathscr{A}$, $\mathscr{A} = \{\alpha = (\alpha^1, \ldots, \alpha^N) | \alpha^i \geq 0 \text{ and } \sum_{i=1}^{N} \alpha^i = 1\}$, and $\mathscr{U}$ is the set of all strategies for all players/agents. The cost values corresponding to the optimal strategies U^* will be $J^1(U^*(\alpha)), \ldots, J^N(U^*(\alpha))$. It is worth noting that although this minimization is over the set of strategies $\mathscr{U}$, the controller parameters (matrices) are in fact being optimized. In other words, the control strategies $\mathscr{U}$ are assumed to be in the form of state feedback and the coefficient matrices are obtained through the above optimization problem.

The strategies obtained from the above minimization as well as the optimal cost values, here referred to as the *solutions*, are functions of the parameter α. Therefore, the Pareto-efficient solution is in general not unique and the set of these solutions, i.e., Pareto frontier, is denoted by $\mathscr{P}$ which is an edge in the space of possible solutions (cost values), i.e., Ξ. It can be shown that in both infinite horizon and finite horizon cases, the Pareto frontier will be a smooth function of α [26]. Due to the nonuniqueness of the Pareto solutions, the next step is to decide how to choose one solution among the set of Pareto solutions (or to choose an α from the set of α's). This solution should be selected according to a certain criterion as the final strategy for the team cooperation problem. For this purpose, we need to solve the bargaining problem as defined below.

Definition 2.4 (Bargaining Problem [26]). In this problem, two or more players/agents have to agree on the choice of certain strategies from a set of solutions while they may have conflicting interests over this set. However, the players/agents recognize that better outcomes may be achieved through cooperation when compared to the noncooperative outcome (known as the threat point). Some of the well-known axiomatic approaches to this problem are the Nash bargaining, the Kalai–Smorodinsky, and the Egalitarian.

Applying any of the above methods to the Pareto-efficient solutions will yield a unique cooperative solution. Due to the interesting properties of the Nash bargaining solution (NBS), such as the symmetry and the Pareto optimality [26], we invoke this method for obtaining a unique solution among the set of Pareto-efficient solutions that are obtained above. The following is a definition for a NBS as presented in [26].

Definition 2.5 (Nash Bargaining Solution (NBS) [26]). In this method, a point in Ξ, denoted by Ξ^N, is selected such that the product of the individual costs from d is maximal ($d = [d^i]^T$ is the threat point or the noncooperative outcome of the agents/players), namely

$$\Xi^N(\Xi,d) = \arg\max_{J\in\Xi} \prod_{i=1}^{N} (d^i - J^i), \quad J = [J^i]^T \in \Xi \text{ with } J \preceq d, \tag{2.40}$$

in which d^i's (the threat point) are the cost values calculated by using the noncooperative solution that is obtained by minimizing the cost in (2.38) individually and constrained to (2.37). It can be shown that the *NBS* is on the Pareto frontier, and therefore the above maximization problem is equivalent to the following problem:

$$\alpha^N = \arg\max_{\alpha} \prod_{i=1}^{N} (d^i - J^i(\alpha, U^*)), \quad J \in \mathscr{P} \text{ with } J \preceq d, \tag{2.41}$$

in which $J = [J^i]^T$ and where J^i's are calculated by using the set of strategies given in (2.39). By solving the maximization problem (2.41), a unique value for the coefficient α can be found.

Remark 2.3. Theorem 6.10 in [26] can be used to determine the relationship that exists between the coefficients α^i, $i = 1,\ldots,N$ and the achievable improvements in the individual costs due to cooperation in the team. According to this theorem the following relationship holds among the values of the cost functions at the NBS, $(J^{1*}(\alpha^*,U^*),\ldots,J^{N*}(\alpha^*,U^*))$, the threat point d, and the optimal weight $\alpha^* = (\alpha^{1*},\ldots,\alpha^{N*})$, i.e.,

$$\alpha^{1*}(d^1 - J^{1*}(\alpha^*,U^*)) = \cdots = \alpha^{N*}(d^N - J^{N*}(\alpha^*,U^*)) \quad \text{or}$$
$$\alpha^{j*} = \frac{\prod_{i\neq j}(d^i - J^{i*}(\alpha^*,U^*))}{\sum_{i=1}^{N}\prod_{k\neq i}(d^k - J^{k*}(\alpha^*,U^*))}. \tag{2.42}$$

The expression in (2.42) describes the kind of cooperation that exists among the players/agents. It shows that if during the team cooperation, i.e., minimization of the team cost, a player/agent has improved its cost more, it will receive a lower weight in the minimization scheme (Pareto solution), whereas the one who has not gained a great improvement as a result of participation in the team cooperation receives a larger weight. Therefore, all the players benefit from the cooperation in almost a similar manner and hence have the incentive to participate in the team cooperation.

Remark 2.4. Selection of the *NBS* is motivated by the fact that this solution enjoys several appealing properties (axioms). As pointed out in [26], in this method each agent does not need to have information about the utility value or the threat point of the other agents. In other words, no "interpersonal comparison" of utility functions is required. Moreover, this solution satisfies four axioms, namely, Pareto optimality, symmetry, independence of irrelevant alternatives, and affine transformation invariance, which are all defined in [26].

2.10 Problem Statement: Consensus in a Team of Multi-agent Systems

The main objective and goal is to make the agents' output, or state vector, converge to a common value, which is either determined by the team members or enforced from outside through a supervisor. The output can be velocity, position, or any other state on which the team should have a consensus on. This implies that $X^i \rightarrow X^j, \forall i, j$. In other words, our objective is to have the team reach to a consensus in the subspace that is spanned by the vector **1**, i.e.,

$$X_{ss} = [(X^1)^T_{ss} \; \dots \; (X^N)^T_{ss}]^T = [1\ 1\ \dots 1]^T \otimes \xi = \mathbf{1} \otimes \xi,$$

where ξ is the final state vector to which the states of all the agents converge to and X_{ss} denotes the steady state vector of the entire team.

Definition 2.6 (Consensus to $\mathscr{S}$ [24]). Let S denote an orthonormal matrix in $R^{Nn\times m}$. The system (2.21) or (2.20) achieves consensus to the subspace $\mathscr{S} = \text{span}\{S\}$ if $\mathscr{S}$ is a minimal set such that for any initial condition, the state $X(t)$ converges to a point in $\mathscr{S}$.

In this book, it is assumed that the desired consensus subspace $\mathscr{S}$ is spanned by the unity vector, i.e., $S = \mathbf{1}$.

Chapter 3
Semi-decentralized Optimal Consensus Strategies

The objectives of this chapter are development and design of controllers for a team of multi-agent systems to accomplish consensus in both the leaderless *(LL)* and the modified leader–follower *(MLF)* architectures in a systematic manner. Towards this end, a semi-decentralized optimal control strategy is designed based on minimization of individual cost functions that are defined for the team members by using local information, only. Furthermore, it is shown that for general linear dynamical models of agents, by appropriate selection of the cost function gains through a set of linear matrix inequalities (LMIs), the team is guaranteed to remain stable and the resulting controllers ensure satisfying the predefined team goal, i.e., cohesive motion of the entire team. It is shown that minimization of the proposed individual cost functions in the infinite horizon case results in a control law which is the well-known "average consensus protocol" [79] or its modified version for both the *LL* and the *MLF* structures. This implies that by using the present framework and methodology the average consensus protocol is derived in a formal manner. In other words, we have introduced a performance index that is minimized by the average consensus protocol through the proposed methodology.

The main feature of our work is to introduce synthesis-based protocols that have the advantage of simultaneously addressing several specifications and constraints while guaranteeing optimality of the solution for a team of multi-agent systems with general linear dynamical models. Modelling of the agents' relative specifications as interaction terms allows and provides a deeper insight into the design of "local" and "global" controllers that address the corresponding local and global specifications. Therefore, interactions among agents due to information flows are represented through the control channels in characterization of dynamical model of each agent as discussed in Chap. 2.

Remark 3.1. In the context of the work presented below, the terms centralized and decentralized can be a source of confusion. Specifically, in our book, a decentralized strategy refers to the case in which the neighbors' information is available for each agent in that set and no exchange of information exists with the other agents that

E. Semsar-Kazerooni and K. Khorasani, *Team Cooperation in a Network of Multi-Vehicle Unmanned Systems: Synthesis of Consensus Algorithms*, DOI 10.1007/978-1-4614-5073-3_3, © Springer Science+Business Media New York 2013

are outside the neighboring set. A centralized strategy refers to a scheme where the entire team information is being made available to each agent.

It should be noted that a preliminary version of the material presented in this chapter has been published in [103, 105–107, 113].

3.1 Optimal Control Problem Formulation

In the following sections, consensus-seeking problem in a network of multi-agent systems is formulated as an optimal control problem.

3.1.1 *Definition of Cost Functions*

To achieve output consensus, the difference among the agents outputs should converge to zero. Towards this end, and due to the connectivity assumption, it suffices to have consensus for the agents in a neighboring set. Hence, the difference between the output of each agent and its neighbors is used as a performance index for each agent. For the leaderless *(LL)* structure (refer to Chap. 2) and the agents' dynamical equation that is given by (2.1) the cost functions for all the agents are defined as given below:

$$\begin{aligned} d^i = & \int_0^T \{ \sum_{j\in N^i} [(Y^i - Y^j)^T Q^{ij} (Y^i - Y^j)] + (u^i)^T R^i u^i \} \mathrm{d}t \\ & + Y^i(T)^T E^i Y^i(T) + (F^i)^T Y^i(T) + G^i \,. \end{aligned} \tag{3.1}$$

If the *MLF* structure is considered, the superscript 1 is used to denote the leader, while the superscripts $i = 2, \ldots, N$ are used to correspond to the variables of the followers. The leader's cost is selected according to

$$\begin{aligned} d^1 = & \int_0^T \{ [(Y^1 - Y^d)^T \Gamma (Y^1 - Y^d)] + \sum_{j\in N^1} [(Y^1 - Y^j)^T Q^{1j} (Y^1 - Y^j)] \\ & + (u^1)^T R^1 u^1 \} \mathrm{d}t + Y^1(T)^T E^1 Y^1(T) + (F^1)^T Y^1(T) + G^1 \,. \end{aligned} \tag{3.2}$$

The cost functions for the followers are defined as in (3.1). In the above definitions, Q^{ij}, Γ, E^i, and R^i are symmetric and positive-definite (PD) matrices, F^i is a vector having a proper dimension, and G^i is a scalar. The variables Y^i, u^i are defined in (2.1), T denotes the time horizon over which the cost optimization is performed, and Y^d is the desired output. The term $Y^i(T)^T E^i Y^i(T) + F^i Y^i(T) + G^i$

denotes the "cost to go." This final value can be either zero or nonzero and its structure is similar to that of the value function. For example, if there is no linear term in the value function, the corresponding linear term in the final value should also be zero. It should be noted that the first term in (3.2) is used to ensure that the leader follows its own command, i.e., $(Y^1 \rightarrow Y^d)$. The second term incorporates the effects of the difference between the output of the leader and the output of its neighbors.

If certain necessary conditions are satisfied, minimization of the above cost functions is guaranteed. This is further discussed in Sects. 3.1.2 and 3.2.1. If these cost functions are minimized then the consensus will be achieved, i.e., all the agents in a neighboring set will reach the same output vector in steady state. Due to the connectivity of the information graph, any neighboring set has at least one common agent with one of the other neighboring sets. Now assume that the agent i has a final output vector Y^{i*}. Any other agent k is either inside or outside its neighboring set. In the former case, $Y^{k*} \rightarrow Y^{i*}$, in which Y^{k*} is the final output of the kth agent. In the latter case, due to the connectivity of the information graph, there is a path between these two agents. Hence, the final output of agent i is passed on to one of its neighbors which in turn passes it to another agent in its own neighboring set until it reaches agent k. Therefore, the two agents will have the same final output. Hence, for the LL structure case all the agents decide on a consensus value Y^c, namely, $Y^{i*} \rightarrow Y^{j*} \rightarrow Y^c, \forall i, j$. The same discussion applies to the *MLF* structure.

Remark 3.2. The cost functions introduced in Eqs. (3.1) and (3.2) are quadratic functions; however, as we will see later due to the dependency of each individual cost function on the outputs of the other agents and due to partial availability of information for individual agents, this quadratic optimal control problem cannot be solved by using conventional *LQR* methods. Specifically, application of these methods to the current problem requires full access to the team information for each agent which is not realistic in practice and is not available. Therefore, in order to determine the minimum value of the above cost function, we have to use the general approach, i.e., to solve the *HJB* equations [9, 26].

Towards this end, let us formulate the consensus-seeking problem and determine what problems may arise in obtaining the solution to the *HJB* equations that are subject to the availability of partial information. This will provide a clear motivation for utilizing our previously introduced interaction terms that are shown to overcome these difficulties. Without loss of generality, we can choose any of the two previously defined structures, namely *LL* or *MLF*. We select the *LL* structure for this purpose and subsequently the consensus problem is solved for both the *LL* and *MLF* structures as shown in Sects. 3.2.1 and 3.2.2.

3.1.2 HJB Equations for the Consensus-Seeking Problem

Assume a team having an *LL* structure where the agent's dynamical equations are governed by (2.1) and the agent's cost function is given as in (3.1). According to the discussion in Chap. 2, the *HJB* equations corresponding to our specific problem are given as follows [9]:

$$
\begin{aligned}
-\frac{\partial V^i}{\partial t}(t,X^i) &= \min_{u\in\mathscr{U}} \Lambda^i(t,X^i,u^i),\\
\Lambda^i &= (X^i)^T Q^i X^i + k_1^i(t,\hat{X}_i^j)X^i + (u^i)^T R^i u^i + k_2^i(t,\hat{X}_i^j)\\
&\quad + \frac{\partial V^i}{\partial X^i}(t,X^i)(A^i X^i + B^i u^i) = -\frac{\partial V^i}{\partial t},\\
V^i(T) &= X^i(T)^T \bar{E}^i X^i(T) + (\bar{F}^i)^T X^i(T) + G^i
\end{aligned}
\tag{3.3}
$$

in which $Q^i = (c^i)^T \sum_{j\in N^i} Q^{ij} c^i$, $\bar{E}^i = (c^i)^T E^i c^i$, $\bar{F}^i = (c^i)^T F^i$, $k_1^i(t,\hat{X}_i^j) = -2\sum_{j\in N^i} (c^j X^j(t))^T Q^{ij} c^i$, $k_2^i(t,\hat{X}_i^j) = \sum_{j\in N^i} (X^j(t))^T (c^j)^T Q^{ij} c^j X^j(t)$, $\hat{X}_i^j$ denotes the set of state vectors of all agents $j \in N^i$, and V^i is a value function to be chosen such that the *PDE* (3.3) is satisfied.

Remark 3.3. Note that (3.3) is a function of not only X^i but also $X^j,\ j \in N^i$. Hence, normally the dynamics of the jth agent should also be considered as a constraint in the *HJB* equations. However, in that case, the solution would be *centralized* in which the control input of each agent depends on the output of *all* the other agents and not necessarily the ones that are in its neighboring set. This is the reason for using only the dynamic constraint of each individual agent in its *HJB* equation and the other agents' states are assumed as time-varying functions.

In view of the above discussion, the minimization of d^i is performed with respect to only $u^i(t,X^i)$ and therefore $u^{i*}(t,X^i) = \arg\{\min_{u^i\in\mathscr{U}^i} \Lambda^i(t,X^i,u^i)\}, i = 1,\ldots,N$. In this case optimality refers to obtaining the minimum value of the cost function (3.1) with the dynamical constraint given by (2.1) while the other agents' control (or states) is treated as being frozen. Therefore, in the context of this chapter, optimality refers to an agent-by-agent optimality rather than the global optimality. Moreover, optimality refers to the optimal performance that is achievable within the specific semi-decentralized structure that is introduced in this book. In other words, the objective is not to show that the semi-decentralized approach is optimal as compared to the centralized case, which indeed it is not. The goal is to obtain an optimal solution for the semi-decentralized approach and this is shown to be achieved if a solution to the proposed optimization problem is obtained.

Discussion on the Existence of a Solution In the above minimization problem, a choice for V^i may be specified as

$$V^i = \frac{1}{2}(X^i)^T K^i(t) X^i + (g^i(t))^T X^i + \gamma^i \tag{3.4}$$

in which K^i, g^i, and γ^i, $i = 1, \ldots, N$, are time-varying parameters to be selected. The reason for this specific selection of V^i is that the optimal control law is of the tracking type, and moreover the terms corresponding to $V^i(T)$ given in (3.3) have similar structure as V^i that is defined in (3.13). Therefore, the *HJB* equations in (3.3) can be solved if K^i and the vector γ^i can be calculated through the following differential equations:

$$-\dot{K}^i = 2Q^i - \frac{1}{2}K^i B^i (R^i)^{-1} (B^i)^T K^i + K^i A^i + (A^i)^T K^i, \; K^i(T) = 2\bar{E}^i$$

$$-\dot{g}^i = -\frac{1}{2}K^i(t) B^i (R^i)^{-1} (B^i)^T g^i + (A^i)^T g^i + k_1^i(t, \hat{X}_i^j)^T, \; g^i(T) = \bar{F}^i .$$

The main issue now is the existence of a solution to the above equations. The first equation is a differential Riccati equation (DRE) and can be solved if certain conditions are satisfied by the matrices A^i, B^i, c^i. Unfortunately, the term $k_1^i(t, \hat{X}_i^j)^T$ in the second equation is a function of the output of the other agents. Although this term is available for the agent i at any time due to being in the neighboring set of agent i, the agent i cannot have its value in advance, which is needed due to the nature of the equations (a two-point boundary value problem), and hence the second equation does not have a solution since Y^j and $j \in N^i$ are not known for the entire interval $[0, T]$. Hence, in this form, there is no solution to g^i. In order to remedy the presence of this unwanted term, the idea of incorporating interaction terms that is introduced in Chap. 2 is invoked. These interaction terms compensate for the effects of Y^j in the cost function of the ith agent. Moreover, this adopted concept allows one to design a "semi-decentralized" controller based on the available information to each agent. This idea is applied and developed further in the subsequent discussions.

Solution of the *HJB* Equations for the Consensus Problem The *HJB* equations given in (3.3) have to be modified to properly address the case when the interaction terms are included in the model as given in (2.6). Moreover, the cost functions (3.1) and (3.2) are modified by replacing u^i, u^1 with u_l^i, u_l^1, respectively, to yield

$$d^i = \int_0^T \{\sum_{j \in N^i} [(Y^i - Y^j)^T Q^{ij} (Y^i - Y^j)] + (u_l^i)^T R^i u_l^i\} \mathrm{d}t$$
$$+ Y^i(T)^T E^i Y^i(T) + (F^i)^T Y^i(T) + G^i , \tag{3.5}$$

$$d^1 = \int_0^T \{[(Y^1 - Y^d)^T \Gamma (Y^1 - Y^d)] + \sum_{j \in N^1} [(Y^1 - Y^j)^T Q^{1j} (Y^1 - Y^j)]$$
$$+ (u_l^1)^T R^1 u_l^1\} \mathrm{d}t + Y^1(T)^T E^1 Y^1(T) + (F^1)^T Y^1(T) + G^1. \tag{3.6}$$

Consequently, the resulting *HJB* equations are modified to

$$\Lambda^i = (X^i)^T Q^i X^i + k_1^i(t,\hat{X}_i^j)X^i + k_2^i(t,\hat{X}_i^j) + (u_l^i)^T R^i u_l^i \\ + \frac{\partial V^i}{\partial X^i}(t,X^i)(A^i X^i + B^i u_l^i(X^i) + B^i u_g^i(Y^j)) = -\frac{\partial V^i}{\partial t}\,. \tag{3.7}$$

The solution to the *HJB* equation in (3.7) yields the following differential equations:

$$-\dot{K}^i = 2Q^i - \frac{1}{2}K^i B^i (R^i)^{-1}(B^i)^T K^i + K^i A^i + (A^i)^T K^i,\; K^i(T) = 2\bar{E}^i \\ -\dot{g}^i = [(A^i)^T - \frac{1}{2}K^i(t)B^i(R^i)^{-1}(B^i)^T]g^i + k_1^i(t,\hat{X}_i^j)^T + K^i(t)B^i u_g^i(Y^j),\; g^i(T) = \bar{F}^i\,. \tag{3.8}$$

The *undesirable* terms, $k_1^i(t,\hat{X}_i^j)$ and $K^i(t)B^i u_g^i(Y^j)$, in the second equation in (3.8) may be canceled out by a suitable selection and design of the interaction terms. Specifically, one solution to consider is to assume that

$$k_1^i(t,\hat{X}_i^j)^T + K^i(t)B^i u_g^i(Y^j) = 0 \Rightarrow \\ -2\sum_{j\in N^i}(c^i)^T Q^{ij} c^j X^j(t) + K^i(t)B^i \sum_{j\in N^i}\mathscr{F}^{ij}Y^j = 0\,. \tag{3.9}$$

This is reduced to the following equation if $K^i(t)B^i$ is a non-singular matrix:

$$\mathscr{F}^{ij} = 2(K^i(t)B^i)^{-1}(c^i)^T Q^{ij}. \tag{3.10}$$

Consequently, one can guarantee that the differential equations in (3.8) have a solution, and therefore a control law may be found to minimize the cost function (3.5). Moreover, using the properties of the Riccati equation and assuming that the pair (A^i,B^i) is reachable and (A^i,Ω^i) is observable, where $Q^{ij} = (\Omega^i)^T\Omega^i$, by putting $\dot{K}^i = 0$ in (3.8), one can conclude that there exists a *PD* solution K^i for the Riccati equation, i.e., the first equation in (3.8) [58]. Therefore, a square and non-singular matrix B^i suffices for the existence of a solution to the second differential equation. In applications related to the ground vehicles, such as in mobile robots, this assumption is quite reasonable.

In the following sections, we discuss two cases, namely a team of multi-agent systems with double integrator dynamical model and a team of agents with linear dynamical model.

3.2 Case I: Team of Multi-agent Systems with Double Integrator Dynamical Model

In this section, we provide solutions to the optimization problem that is defined in the previous section by assuming that the dynamical behavior of multi-agent systems is governed by a double integrator model.

3.2.1 *Consensus Problem in a Leaderless Multi-vehicle Team*

Consider a team of multi-agent systems with a structure that is similar to Fig. 2.1a, with the agents' dynamics as given by (2.20) and with the cost function as specified in (3.5) by replacing Y^i with v^i. The matrix Q^{ij} should be selected such that the pair $(0, \sqrt{Q^{ij}})$ is observable. The corresponding *HJB* equations are then similar to (3.3) by assuming that $X^i = [r^i \; v^i]^T$ and Λ^i is defined as follows:

$$\begin{aligned}\Lambda^i(t,X^i,u^i) &= \sum_{j\in N^i} (v^i - v^j)^T Q^{ij}(v^i - v^j) + (u_l^i)^T R^i u_l^i \\ &\quad + \frac{\partial V^i}{\partial v^i}(t,X^i)\left(u_l^i + \sum_{j\in N^i} \mathscr{F}^{ij} v^j\right) + \frac{\partial V^i}{\partial r^i}(t,X^i)v^i \,, \end{aligned} \tag{3.11}$$

in which V^i at any time t has the following value:

$$\begin{aligned}V^i(t) &= \min_{u_l^i} \int_t^T \left[\sum_{j\in N^i} \left[(v^i - v^j)^T Q^{ij}(v^i - v^j)\right] + (u_l^i)^T R^i u_l^i\right] \mathrm{d}t \\ &\quad + v^i(T)^T E^i v^i(T) + (F^i)^T v^i(T) + G^i \,, \end{aligned} \tag{3.12}$$

and a choice of V^i may be specified as

$$V^i = \frac{1}{2}(v^i)^T K^i(t) v^i + \gamma^i(t) \,, \tag{3.13}$$

in which K^i, γ^i $i = 1,\ldots,N$ are time-varying parameters to be defined.

Existence of a Solution To guarantee existence of a solution to the above minimization problem, controllability of the open-loop system should be verified [58], i.e., the pair (A^i, B^i) in (2.6) should be controllable. According to the representation of the system that is given in (2.20), the required matrices in this case are $A^i = \left[\begin{smallmatrix} 0_{2\times2} & I_{2\times2} \\ 0_{2\times2} & 0_{2\times2}\end{smallmatrix}\right]$ and $B^i = \left[\begin{smallmatrix} 0_{2\times2} \\ I_{2\times2}\end{smallmatrix}\right]$. However, since only the dynamics of v^i appear in the cost function (3.5), and the dynamics of r^i do not have any effect on the v^i dynamics, to verify the controllability condition, we need to only consider the dynamics of v^i. That is, we have

$$A^i = 0_{2\times 2},\ B^i = I_{2\times 2},\ i = 1,\ldots,N\,. \tag{3.14}$$

This pair is controllable and therefore guarantees the existence of a solution to the corresponding *HJB* equations. Furthermore, based on the definition of the optimality provided previously, the presence of the term $\sum_{j\in N^i}\mathscr{F}^{ij}v^j$ in (2.20) does not have any effect on the matrices that are involved in the controllability condition. The above reasoning applies to both the *LL* and the *MLF* structures with agents integrator dynamics.

In order to have a solution to the *PDE*s in (3.11), it can be shown that K^i and the interaction terms should be computed according to the following lemma.

Lemma 3.1. *Assume that an* LL *team of mobile robots whose dynamics are governed by the double integrator equations operates subject to the interactions among the vehicles based on the neighboring sets as given by (2.20) for an undirected and connected network structure. The interaction coefficient terms and the control law proposed below will minimize the cost function (3.5) and also guarantee alignment of the team of vehicles (consensus over the velocity), where*

$$u_g^{i*} = \sum_{j\in N^i}\mathscr{F}^{ij}v^j,\ \mathscr{F}^{ij} = 2K^i(t)^{-1}Q^{ij},\ \forall i,j = 1,\ldots,N\,, \tag{3.15}$$

and

$$u_l^{i*} = -\frac{1}{2}(R^i)^{-1}K^i(t)v^i,\ i = 1,\ldots,N\,, \tag{3.16}$$

where u_g^{i*}, u_l^{i*} designate the optimal values of u_g^i, u_l^i, respectively, and K^i satisfies the following DRE*:*

$$-\dot{K}^i = 2|N^i|Q^{ij} - \frac{1}{2}K^i(R^i)^{-1}K^i,\ K^i(T) = 2E,\ i = 1,\ldots,N\,. \tag{3.17}$$

Proof. The details are given in Appendix A. □

Remark 3.4. It can be seen from (3.16) that the control law u_l^{i*} depends *only* on the state of the agent *i*, and the term u_g^{i*} ensures that the effects of the neighboring agents on the dynamics of agent *i* are present. This provides a semi-decentralized control protocol according to the information exchanges that are within the neighboring sets.

The next theorem provides important properties of our proposed control strategy.

Theorem 3.1. ***(a) Consensus Protocol:*** *For the team of mobile robots that is described in Lemma 3.1, and under the condition of infinite horizon scenario (i.e., $T \longrightarrow \infty$), and for an undirected and connected network structure, the combined control law reduces to the following well-known average consensus (agreement) protocol, i.e.,*

$$u^{i*}(v^i, v^j) = u_l^{i*}(v^i) + u_g^{i*}(v^j) = \Gamma^i (v^i - \frac{\sum_{j \in N^i} v^j}{|N^i|}),$$

$$\Gamma^i = -\frac{1}{2}(R^i)^{-1} K^i, \ i = 1, \ldots, N. \tag{3.18}$$

(b) Stability: *The above protocol furthermore guarantees consensus on a constant common value, v^c, in a globally asymptotic manner, i.e.,*

$$v^i \to v^j \to v^c, \ \ as\, t \to \infty, \ \ \forall i, j = 1, \ldots, N. \tag{3.19}$$

(c) Consensus Value: *The consensus value that is achieved by the team will be the following:*

$$v^c = (w_1 w_1^T + w_2 w_2^T) Avg(v(0)), \tag{3.20}$$

in which w_1 and w_2 are the eigenvectors of the matrix Γ^i, v^c is the consensus vector and $v(0)$ is the concatenation vector of the initial velocities of agents, i.e., $v(0) = [(v^1(0))^T \ \ldots \ (v^N(0))^T]^T$.

Proof. The details are provided in Appendix A. □

Remark 3.5. It follows readily from Theorem 3.1 that in the infinite horizon case the control law obtained by the proposed method is the well-known average consensus protocol that is introduced in [79].

3.2.2 Consensus Problem in an MLF Multi-vehicle Team

Now consider a team of multi-agent systems with an *MLF* structure as described in Fig. 2.1b. The dynamics of the agents are given by (2.20) and the cost functions for the leader and the followers as specified according to Eqs. (3.6) and (3.5), respectively, by replacing Y^i with v^i. The *HJB* equations are reduced to the following:

$$-\frac{\partial V^i}{\partial t}(t, X^i) = \min_{u \in \mathcal{U}} \Lambda^i(t, X^i, u^i), \tag{3.21}$$

$$\Lambda^1(t, X^1, u^1) = \sum_{j \in N^1} (v^1 - v^j)^T Q^{1j} (v^1 - v^j) + (v^1 - v^d)^T \Gamma (v^1 - v^d)$$

$$+ (u_l^1)^T R^1 u_l^1 + \frac{\partial V^1}{\partial v^1}(u_l^1 + \sum_{j \in N^1} \mathcal{F}^{1j} v^j), \tag{3.22}$$

$$\Lambda^i(t,X^i,u^i) = \sum_{j\in N^i} (v^i - v^j)^T Q^{ij}(v^i - v^j) + (u_l^i)^T R^i u_l^i$$

$$+\frac{\partial V^i}{\partial v^i}(u_l^i + \sum_{j\in N^i} \mathscr{F}^{ij} v^j),\ i = 2,\ldots,N\ , \tag{3.23}$$

where a choice of V^i and V^1 is specified as

$$\begin{aligned} V^1 &= \frac{1}{2}(v^1)^T K^1(t) v^1 + (g^1(t))^T v^1 + \gamma^1(t), \\ V^i &= \frac{1}{2}(v^i)^T K^i(t) v^i + \gamma^i(t),\ i = 2,\ldots,N\ , \end{aligned} \tag{3.24}$$

in which K^i and γ^i, $i = 1,\ldots,N$, and g^1 are time-varying parameters to be chosen.

In order to obtain a solution to the *PDE*s in Eqs. (3.21)–(3.23), u^i and the interaction terms $\mathscr{F}^{ij}$ should be selected according to the following lemma.

Lemma 3.2. *Assume a team of mobile robots whose dynamics are governed by the double integrator equations as given in (2.20) and having an* MLF *structure. The interaction terms and the control laws proposed below will provide a solution to the* HJB *equations in Eqs. (3.21)–(3.23) and will simultaneously minimize the cost functions (3.5) and (3.6) to guarantee the alignment of the vehicles, where*

$$u_g^{i*} = \sum_{j\in N^i} \mathscr{F}^{ij} v^j = \sum_{j\in N^i} 2(K^i)^{-1} Q^{ij} v^j,\ \forall i\ , \tag{3.25}$$

$$u_l^{i*} = -\frac{1}{2}(R^i)^{-1} K^i(t) v^i,\ i = 2,\ldots,N\ , \tag{3.26}$$

$$u_l^{1*} = -\frac{1}{2}(R^1)^{-1}(K^1(t)v^1 + g^1(t))\ , \tag{3.27}$$

and where the leader's parameter g^1 *and the* DREs *for determining* K^i *satisfy*

$$-\dot{K}^i = 2|N^i|Q^{ij} - \frac{1}{2}K^i(R^i)^{-1}K^i,\ K^i(T) = 2E^i,\ i = 2,\ldots,N, \tag{3.28}$$

$$-\dot{K}^1 = 2(|N^1|Q^{1j} + \Gamma) - \frac{1}{2}K^1(R^1)^{-1}K^1,\ K^1(T) = 2E^1, \tag{3.29}$$

$$\dot{g}^1 = 2\Gamma v^d(t) + \frac{1}{2}K^1(R^1)^{-1}g^1,\ g^1(T) = F^1\ . \tag{3.30}$$

Proof. The details are provided in Appendix A. □

The discussions on the existence of a solution are similar to those presented in Sect. 3.2.1. The next theorem provides important properties of the proposed control strategy.

Theorem 3.2. ***(a) Modified Consensus Protocol:*** *For the team of vehicles that is described in Lemma 3.2, and associated with the infinite horizon scenario (i.e.,* $T \longrightarrow \infty$*), the combined control law reduces to the modified agreement protocol for the* MLF *structure. The protocol for a follower is given by*

$$u^{i*}(v^i, v^j) = u_l^{i*}(v^i) + u_g^{i*}(v^j) = \Gamma^i(v^i - \frac{\sum_{j\in N^i} v^j}{|N^i|}), \tag{3.31}$$

and for the leader is given by

$$u^{1*}(v^1, v^j) = \Gamma^1(v^1 - \frac{\sum_{j\in N^1} v^j}{|N^1|}) + \beta^1(v^1 - v^d), \tag{3.32}$$

$$\Gamma^i = -2(K^i)^{-1}|N^i|Q^{ij},\ i = 1,\ldots,N,\ \ \beta^1 = -2(K^1)^{-1}\Gamma\,. \tag{3.33}$$

(b) Stability: *The above protocol is stabilizing, i.e., the error dynamics of the entire team is asymptotically stable, implying that*

$$e^i(t) = v^i(t) - v^d \to 0 \ \text{as}\ t \to \infty,\ \ i = 1,\ldots,N\,. \tag{3.34}$$

Proof. The details are provided in Appendix A. □

It is seen that the second term of the leader's control input in (3.32) guarantees the command tracking even if the first term is not included. However, in Chap. 4, it will be shown that the first part keeps and maintains the team cohesion and guarantees that none of the followers are *lost* without affecting the others and specifically the leader's behavior.

3.3 Case II: Team of Multi-agent Systems with Linear Dynamical Model

In this section, we provide solutions to the optimization problem that was defined previously, assuming that the dynamical behavior of multi-agent systems is governed by a linear model.

3.3.1 Consensus Problem in an MLF Multi-vehicle Team

Assume that the agents are ground vehicles that are governed by linear dynamics as given in (2.21). Now consider an *MLF* structure as shown in Fig. 2.1b and the cost functions for the followers and the leader as specified according to Eqs. (3.5) and (3.6) by replacing Y^i with v^i, respectively. The *HJB* equations will consequently reduce to the following:

$$-\frac{\partial V^i}{\partial t}(t,v^i) = \min_{u^i \in \mathscr{U}^i} \Lambda^i(t,v^i,u^i) , \tag{3.35}$$

$$\begin{aligned}
\Lambda^1(t,v^1,u^1) &= \sum_{j\in N^1} (v^1 - v^j)^T Q^{1j}(v^1 - v^j) + (v^1 - v^d)^T \Gamma (v^1 - v^d) \\
&\quad + (u_l^1)^T R^1 u_l^1 + \frac{\partial V^1}{\partial v^1}(t,v^1)(A^1 v^1 + B^1(u_g^1 + u_l^1)), \\
V^1(T,v^1) &= v^1(T)^T E^1 v^1(T) + (F^1)^T v^1(T) + G^1(T) ,
\end{aligned} \tag{3.36}$$

$$\begin{aligned}
\Lambda^i(t,v^i,u^i) &= \sum_{j\in N^i} (v^i - v^j)^T Q^{ij}(v^i - v^j) + (u_l^i)^T R^i u_l^i \\
&\quad + \frac{\partial V^i}{\partial v^i}(t,v^i)(A^i v^i + B^i(u_g^i + u_l^i)), \\
V^i(T,v^i) &= v^i(T)^T E^i v^i(T) + (F^i)^T v^i(T) + G^i(T),\ i = 2,\dots,N .
\end{aligned} \tag{3.37}$$

Therefore, a choice of V^i for the followers and the leader may be specified according to

$$V^i = \frac{1}{2}(v^i)^T K^i(t) v^i + \gamma^i(t),\ i = 2,\dots,N , \tag{3.38}$$

$$V^1 = \frac{1}{2}(v^1)^T K^1(t) v^1 + (g^1(t))^T v^1 + \gamma^1(t) , \tag{3.39}$$

where K^i, γ^i, and g^1 are the time-varying parameters that are to be specified.

Existence of a Solution To guarantee existence of a solution to the above minimization problem the notion of reachability should be verified for the open-loop system [58]. Based on the representation of the system (2.21), and given that only the dynamics of v^i appears in the cost function (3.5), to verify the reachability condition, we need to only consider the dynamics of v^i. Furthermore, based on the definition of the optimality given previously, where the other agents' dynamics are considered as time-varying functions only, the presence of the term $B^i \sum_{j\in N^i} \mathscr{F}^{ij} Y^j$ in (2.21) does not have any effect on the matrices that are involved in the reachability condition. Therefore, due to the non-singularity of the matrix B^i, the pair (A^i, B^i) is always reachable, and therefore existence of a solution is always guaranteed.

Solution of the corresponding *HJB* equations is now provided in the following lemma.

Lemma 3.3. *Assume a team of multi-agent systems whose dynamics are governed by (2.21) with the pair (A^i,Ω^i) being observable, where $Q^{ij}=(\Omega^i)^T\Omega^i$, for an* MLF *structure. The leader is aware of the desired command specifications and requirements while the followers operate subject to interactions among the agents based on the neighboring sets. The interaction terms and the control law proposed below will provide a solution to the* HJB *equations in Eqs. (3.36) and (3.37) and simultaneously minimize the cost function (3.5) for the followers and the cost function (3.6) for the leader and guarantee a consensus achievement with the consensus state of $v^i=v^d$, $\forall i$, where*

$$u_g^{i*}=\sum_{j\in N^i}\mathscr{F}^{ij}v^j=\sum_{j\in N^i}2(K^iB^i)^{-1}Q^{ij}v^j,\ i=1,\ldots,N\,, \tag{3.40}$$

$$u_l^{i*}=-\frac{1}{2}(R^i)^{-1}(B^i)^TK^i(t)v^i,\ i=2,\ldots,N\,, \tag{3.41}$$

$$u_l^{1*}=-\frac{1}{2}(R^1)^{-1}(B^1)^T(K^1(t)v^1+g^1(t))\,, \tag{3.42}$$

and the leader's parameter g^1 and the Riccati equations for determining K^i satisfy

$$\begin{aligned}-\dot{K^i}&=2|N^i|Q^{ij}-\frac{1}{2}K^iB^i(R^i)^{-1}(B^i)^TK^i+(A^i)^TK^i+K^iA^i,\\ K^i(T)&=2E^i,\ i=2,\ldots,N\,,\end{aligned} \tag{3.43}$$

$$\begin{aligned}-\dot{K^1}&=2(|N^1|Q^{1j}+\Gamma)+(A^1)^TK^1+K^1A^1-\frac{1}{2}K^1B^1(R^1)^{-1}(B^1)^TK^1,\\ K^1(T)&=2E^1\,,\end{aligned} \tag{3.44}$$

$$\dot{g^1}=2\Gamma v^d(t)+(\frac{1}{2}K^1B^1(R^1)^{-1}(B^1)^T-(A^1)^T)g^1,\ g^1(T)=F^1\,. \tag{3.45}$$

Proof. The details are provided in Appendix A. □

The next theorem provides the stability property of the closed-loop team of multi-agent systems as well as the behavior of the team in achieving consensus.

Theorem 3.3. ***(a) Modified Consensus Protocol:*** *For the team of multi-agent systems described in Lemma 3.3, and associated with the infinite horizon scenario (i.e., when $T\longrightarrow\infty$), the combined control law reduces to a modified average consensus protocol (agreement protocol) for the* MLF *structure. The protocol for the followers and the leader are given by*

$$u^{i*}(v^i,v^j)=u_l^{i*}(v^i)+u_g^{i*}(v^j)=\Gamma^i(v^i-\frac{\sum_{j\in N^i}v^j}{|N^i|})+\beta^iv^i,\ i=2,\ldots,N\,, \tag{3.46}$$

$$u^{1*} = u_l^{1*}(v^1) + u_g^{1*}(v^j) = \Gamma^1(v^1 - \frac{\Sigma_{j \in N^1} v^j}{|N^1|}) + \alpha^1(v^1 - v^d)$$
$$+\beta^1 v^1 - (K^1 B^1)^{-1}(A^1)^T g^1 \,, \tag{3.47}$$

in which $\alpha^1 = -2(K^1B^1)^{-1}\Gamma$, $\Gamma^i = -2(K^iB^i)^{-1}|N^i|Q^{ij}$, *and for* $\forall i$, $\beta^i = -(K^iB^i)^{-1}(K^iA^i + (A^i)^T K^i)$.

(b) ***Stability:*** *The above protocol is stabilizing, i.e., the error dynamics of the entire team is asymptotically stable, implying that*

$$e^i = v^i - v^d \to 0 \;\; as\, t \to \infty, \;\; i = 1, \ldots, N \tag{3.48}$$

if the parameters Γ *and* R^1 *in the cost function (3.6) and the matrix* K^1 *in (3.44) are determined appropriately, and moreover the following set of* LMIs *in terms of* Γ, K^1, Z^1 ($R^1 = -\frac{1}{2}(B^1)^T K^1 (Z^1)^{-1} K^1 B^1$) *is satisfied:*

$$\begin{cases} \Upsilon + \Upsilon^T > 0, \;\; \Upsilon = 2L \otimes Q^{ij} + A^T K + 2G \\ (A^1)^T K^1 + K^1 A^1 + Z^1 + 2(|N^1| Q^{1j} + \Gamma) = 0 \\ Z^1 = -\frac{1}{2} K^1 B^1 (R^1)^{-1} (B^1)^T K^1, \; \Gamma > 0, \; K^1 > 0, \; Z^1 < 0 \end{cases} \tag{3.49}$$

where $K = Diag\{K^i, i = 1, \ldots, N\}$, $A = Diag\{A^i, i = 1, \ldots, N\}$, *and* $G = Diag\{\Gamma, 0, \ldots, 0\}$.

Proof. The details are provided in Appendix A. □

Remark 3.6. It follows readily from Theorem 3.3 that in the infinite horizon case the control law obtained by the proposed methodology is a modified version of the well-known average consensus protocol [79].

3.3.2 Consensus Problem in an LL Multi-vehicle Team

In this case, the multi-agent system dynamics are governed by (2.21), and the cost function for all the agents is given by (3.5) in the *LL* structure similar to Fig. 2.1a. The HJB equations defined previously will reduce to Eqs. (3.35) and (3.37). Similar to the previous discussions, and in order to have a decentralized solution, the minimization is performed with respect to v^i only, and a choice for V^i may be specified as in (3.38). The existence of a solution to this problem follows along the similar lines as those invoked for the *MLF* structure. As a matter of fact, these attributes are invariant under topological changes, i.e., they are valid for both the leaderless or the leader–follower architectures.

Based on the choice of V^i as in (3.38), and in order to have a solution to the PDEs in Eqs. (3.35) and (3.37), it can be shown that K^i and the interaction terms should be computed according to the following lemma.

Lemma 3.4. *Assume a team of vehicles whose dynamics are governed by (2.21) for the* LL *structure. The team operates subject to interactions among vehicles based on the neighboring sets. The interaction terms and the control law proposed below will provide a solution to the* HJB *equations in Eqs. (3.35) and (3.37) and simultaneously minimize the cost function (3.5) and guarantee consensus achievement with the consensus state of* $Y^i = v^i = v^c$, $\forall i$, *with*

$$u_g^{i*} = \sum_{j \in N^i} \mathscr{F}^{ij} v^j = \sum_{j \in N^i} 2(K^i B^i)^{-1} Q^{ij} v^j, \; i = 1, \ldots, N, \tag{3.50}$$

$$u_l^{i*} = -\frac{1}{2}(R^i)^{-1}(B^i)^T K^i(t) v^i, \; i = 1, \ldots, N, \tag{3.51}$$

$$-\dot{K}^i = 2|N^i|Q^{ij} - \frac{1}{2}K^i B^i (R^i)^{-1}(B^i)^T K^i + (A^i)^T K^i + K^i A^i, \; K^i(T) = 2E^i. \tag{3.52}$$

Proof. It follows along the similar lines as in the proof of Lemma 3.3, and is therefore omitted. □

The next theorem provides the characteristics and features of the proposed control strategy in terms of the already well-known behavior of cooperative teams as well as the requirement of guaranteeing team stability.

Theorem 3.4. ***(a) Consensus Protocol:*** *For the team of vehicles that is described in Lemma 3.4, and associated with the infinite horizon scenario (i.e.,* $T \longrightarrow \infty$*), the combined control law reduces to the modified average consensus protocol (agreement protocol) for the* LL *structure. The protocol for the agents is given by*

$$u^{i*}(v^i, v^j) = u_l^{i*}(v^i) + u_g^{i*}(v^j) = \Gamma^i\left(v^i - \frac{\sum_{j \in N^i} v^j}{|N^i|}\right) + \beta^i v^i, \tag{3.53}$$

where $\Gamma^i = -2(K^i B^i)^{-1}|N^i|Q^{ij}$ *and* $\beta^i = -(K^i B^i)^{-1}(K^i A^i + (A^i)^T K^i)$, $i = 1, \ldots, N$.

(b) Stability: *The above protocol furthermore guarantees consensus on a constant common value,* v^c*, in a globally asymptotic manner, i.e.,*

$$v^i \to v^j \to v^c, \;\; \text{as } t \to \infty, \;\; \forall i, j = 1, \ldots, N, \tag{3.54}$$

if A^i *satisfies the condition* $(A^i)^T K^i + K^i A^i \geq 0$ *with* A^i *having at least one zero eigenvalue.*

(c) Consensus Value: *The consensus value achieved by the team, i.e.,* v^c*, is in the null space of* $(A^i)^T K^i$.

Proof. The details are provided in Appendix A. □

3.4 Simulation Results

In the following sections we provide simulation results for implementing and applying our proposed methods that are developed in the previous sections to a team of multi-agent systems.

3.4.1 *Double Integrator Dynamical Model*

In this section, simulation results are presented for both the finite and the infinite horizon scenarios for the *LL* and the *MLF* structures. To obtain numerical solutions to the *DRE* in the finite horizon case, the backward differentiation formula (BDF) that is described in [14] is used. The simulations are conducted for a team of four mobile robots with dynamical equations that are given in (2.20). Without loss of generality, the topology is assumed to be a ring topology. The objective of the team is to ensure that all the agents have the same velocity in the steady state. Results shown below are conducted to capture the average behavior of the proposed control strategies through Monte-Carlo simulations. The *average* team response due to 30 different *randomly* selected initial conditions is presented.

In Fig. 3.1a, b the x- and y-components of v^i, i.e., v^i_x, v^i_y are shown for the *LL* structure in the infinite horizon case. Figure 3.1c illustrates the actual average path trajectories generated by the vehicles in the $x-y$ plane. It may be observed that the vehicles are aligned and move together after the transients have died out. Figure 3.2a–c depicts the same trajectories but now for the finite horizon scenario.

For the *MLF* structure, simulations are conducted for the same configuration as above; however, the objective is to ensure that the team members have the same velocity as the desired value specified by the leader, i.e., v^d. In Fig. 3.3a, b, the x- and y-components of v^i, i.e., v^i_x, v^i_y, are shown for the finite horizon scenario. The presence of transients in the final state of the variables is due to the finite horizon formulation of the optimal control problem (a two-point boundary value problem). The desired velocity (leader command) is chosen as $v^d = [3 \;\; 4]^T (m/s)$. Figure 3.3c shows the actual path trajectories that are generated by the vehicles in the $x-y$ plane. Figure 3.4a–c depicts the same trajectories but now for the infinite horizon scenario. In all the simulations the parameters selected are $Q^{ij} = \left[\begin{smallmatrix} 1 & 0 \\ 0 & 3 \end{smallmatrix}\right], R^i = I_{2\times 2}, |N^i| = 2$, $E^i = 0.5 I_{2\times 2}$, and $\Gamma = \left[\begin{smallmatrix} 10 & 0 \\ 0 & 40 \end{smallmatrix}\right]$, and the random initial conditions, i.e., $X^i_0 = [(r^i(0))^T \;\; (v^i(0))^T]^T$, $\forall i$, for the Monte-Carlo simulations are considered as $X^1_0 = [r(0,1) \;\; r(0,1) \;\; r(-5,0) \;\; r(-6,-1)]^T$, $X^2_0 = [r(1,2) \;\; r(1,2) \;\; r(0,5) \;\; r(-5,5)]^T$, $X^3_0 = [r(2,3) \;\; r(2,3) \;\; r(5,10) \;\; r(1,11)]^T$, and $X^4_0 = [r(3,4) \;\; r(3,4) \;\; r(-10,-5) \;\; r(-6,4)]^T$ for the *LL* structure and $X^1_0 = [r(0,15) \; r(0,25) \; r(-5,0) \; r(-6,-1)]^T, X^2_0 = [r(15,30) \, r(25,50) \, r(0,5) r(-5,5)]^T$, $X^3_0 = [r(30,45) \;\; r(50,75) \;\; r(5,10) \;\; r(1,11)]^T$, and $X^4_0 = [r(45,60) \;\; r(75,100) \;\; r(-10,-5) \;\; r(-6,4)]^T$ for the *MLF* structure, where $r(x,y)$ designates a random

Table 3.1 The mean performance index corresponding to different team structures and control design assumptions

Average performance index for the Monte-Carlo simulations				
Team structure	Agent 1	Agent 2	Agent 3	Agent 4
LL (FH)	2,488.2	2,163.2	4,008.5	4,510.7
LL (IH)	1,963.7	1,716.1	3,794.3	4,198.4
MLF (FH)	231,990	16,964	20,096	19,287
MLF (IH)	222,820	15,899	17,264	16,822

FH and IH denote the finite and the infinite horizon scenarios, respectively

variable in the interval $[x,y]$. Also, for the *LL* structure, the time horizon is selected as $T=6$ and for the *MLF* structure $T=15$.

In order to provide a better insight into the controller performance for different cases that are considered above, the average cost values that are obtained by running the Monte-Carlo simulations corresponding to each team structure and design assumptions are provided in Table 3.1. One can conclude that the finite horizon design scenario results in a higher average cost as compared to the infinite horizon design scenario. Note that the average velocities corresponding to the *LL* structure are $[-1.46\ 0.108]^T$ for the infinite horizon scenario (and are $[-1.35\ 0.422]^T$ for the results corresponding to the finite horizon scenario).

3.4.2 Linear Dynamical Model

In this section, simulation results are presented for an infinite horizon scenario for the *LL* and the *MLF* structures associated with a linear model of multi-agent systems [as given in (2.21)]. The simulations are conducted for a team of four vehicles where the objective for the team is to ensure that all the agents have the same velocity in the steady state. Without loss of generality, and for the sake of only simulations, the topology is assumed to be a ring. For the *MLF* structure, the objective is to ensure that the team members have the same velocity as the desired value specified by the leader, i.e., $v^d(t)$. In Fig. 3.5a, b, the x- and y-components of v^i are shown for $i=1,\dots,4$. Figure 3.5c depicts the actual path trajectories generated by the vehicles in the $x-y$ plane. It may be concluded that vehicles are aligned and move together with the same desired velocity as the command provided by the leader. For simulations, the command is assumed to be a pulse-like signal with a duration of $0.4\,sec$ and its value switches between $v^d=[3\ \ 4]^T$ and $v^d=[5\ \ -1]^T$. The initial state of the vehicles, i.e., $X_0^i=[(r^i(0))^T\ \ (v^i(0))^T]^T$, $\forall i$, are selected at $X_0^1=[0.6\ \ 1\ \ 5\ \ 3]^T$, $X_0^2=[2\ \ 1\ \ -5\ \ -4]^T$, $X_0^3=[0.4\ \ 3\ \ -1\ \ -2]^T$, and $X_0^4=[2\ \ 0\ \ 3\ \ 4]^T$, and the other simulation parameters are chosen as $Q^{ij}=100I_{2\times 2}, |N^i|=2$, and for $i=2,3,4$, $R^i=0.01I_{2\times 2}$. The parameters corresponding to the agent model are chosen as $\bar{A}^i=\left(\begin{smallmatrix}1&0\\2&6\end{smallmatrix}\right), A^i=\left(\begin{smallmatrix}-1&0\\0&2\end{smallmatrix}\right)$, and $B^i=\left(\begin{smallmatrix}1&1\\2&6\end{smallmatrix}\right)$. The matrices R^1, K^1, Γ are obtained by

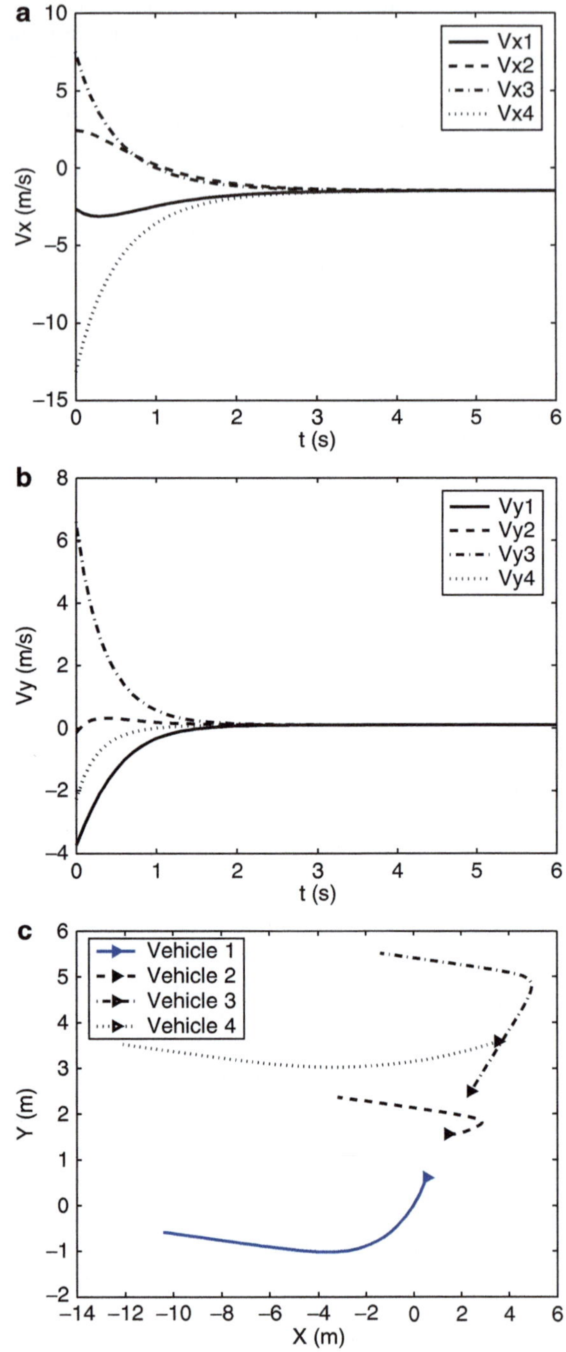

Fig. 3.1 **(a)** The x-component of the average velocity profile, **(b)** the y-component of the average velocity profile, and **(c)** the $x-y$ path trajectories of the *LL* team of four agents (Monte-Carlo simulation runs) resulting from the optimal control strategy in the infinite horizon scenario

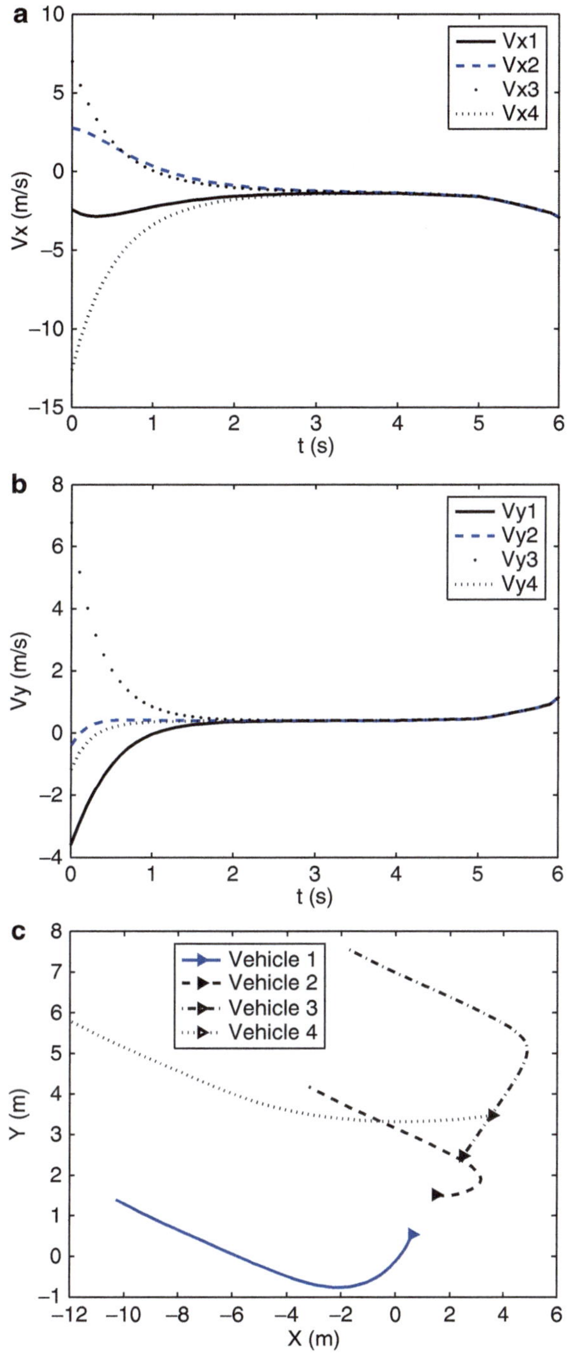

Fig. 3.2 **(a)** The x-component of the average velocity profile, **(b)** the y-component of the average velocity profile, and **(c)** the $x-y$ path trajectories of the LL team of four agents (Monte-Carlo simulation runs) resulting from the optimal control strategy in the finite horizon scenario

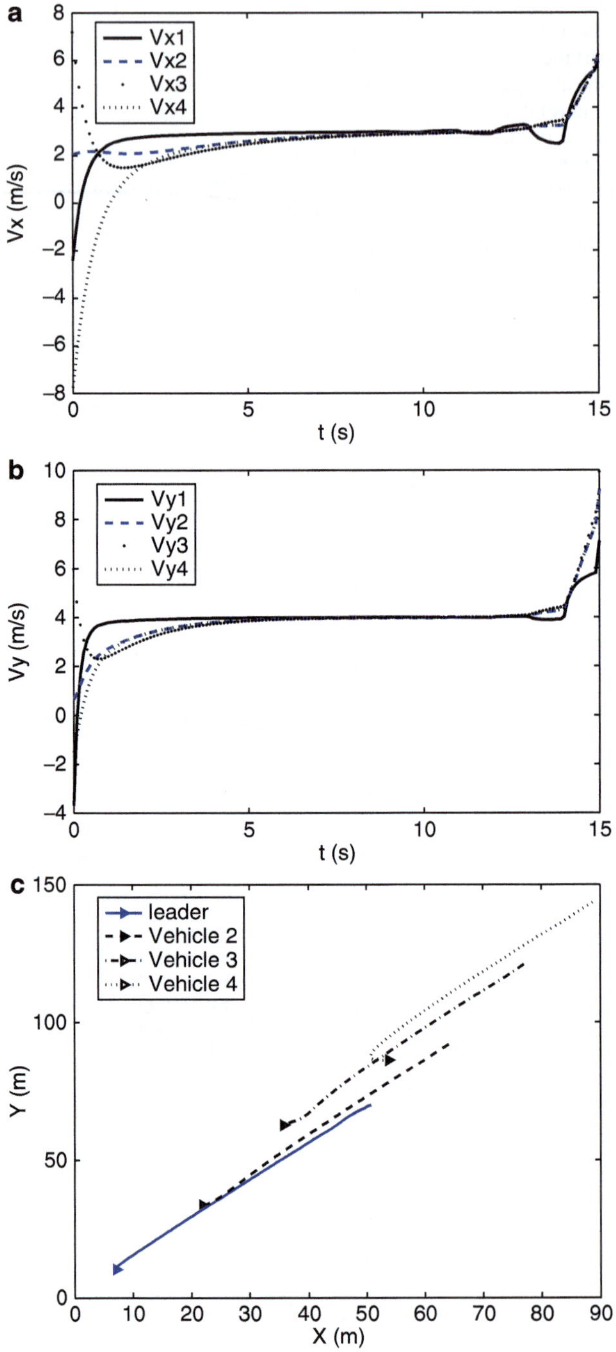

Fig. 3.3 **(a)** The x-component of the average velocity profile, **(b)** the y-component of the average velocity profile, and **(c)** the $x-y$ path trajectories of the *MLF* team of four agents (Monte-Carlo simulation runs) resulting from the optimal control strategy in the finite horizon scenario

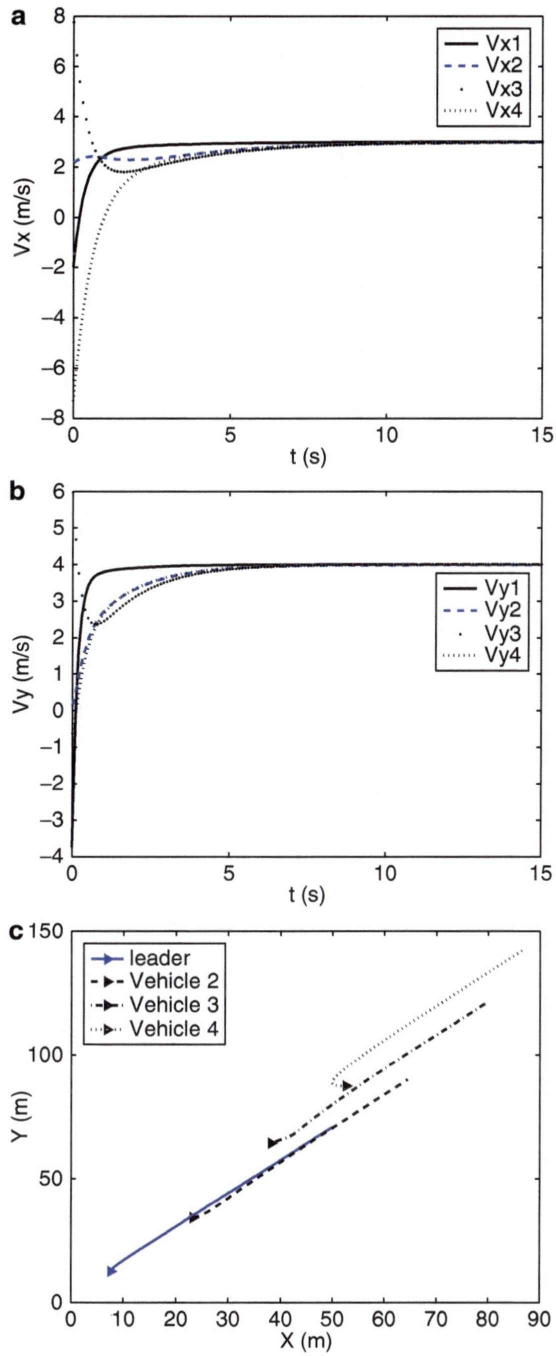

Fig. 3.4 **(a)** The x-component of the average velocity profile, **(b)** the y-component of the average velocity profile, and **(c)** the $x-y$ path trajectories of the *MLF* team of four agents (Monte-Carlo simulation runs) resulting from the optimal control strategy in the infinite horizon scenario

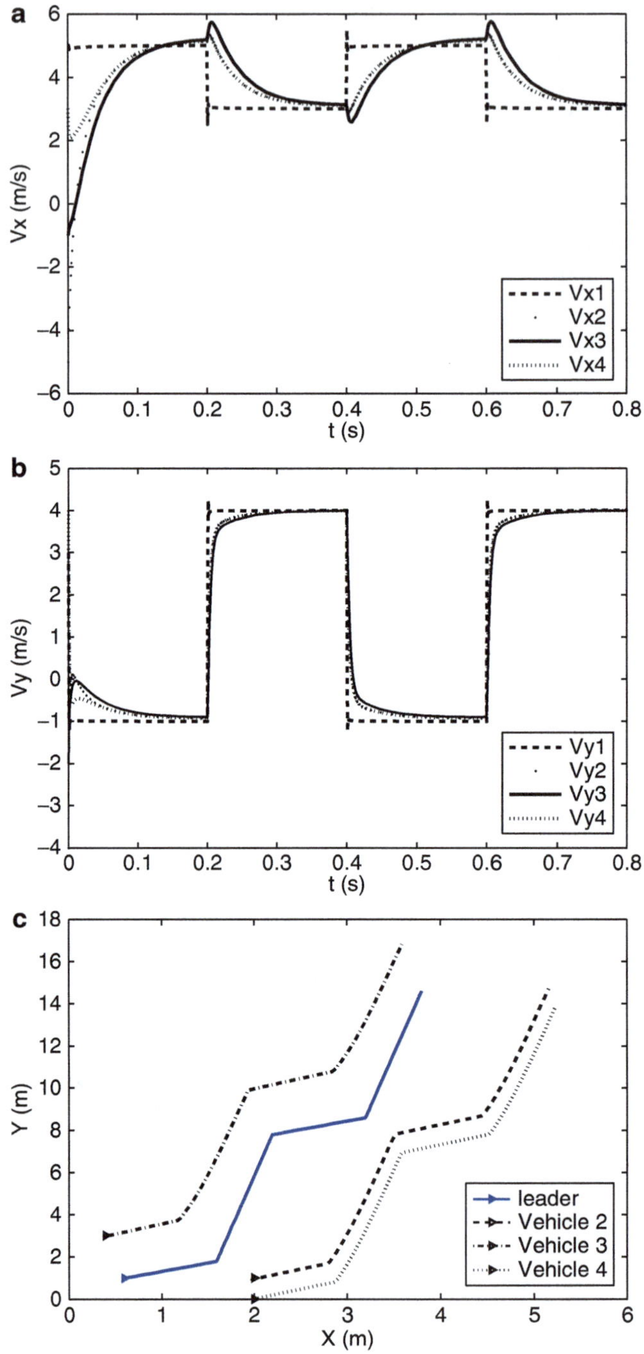

Fig. 3.5 **(a)** The x-component of the velocity profile, **(b)** the y-component of the velocity profile, and **(c)** The $x-y$ path trajectories of the *MLF* team of four agents with linear dynamical model resulting from the optimal control strategy in the infinite horizon scenario

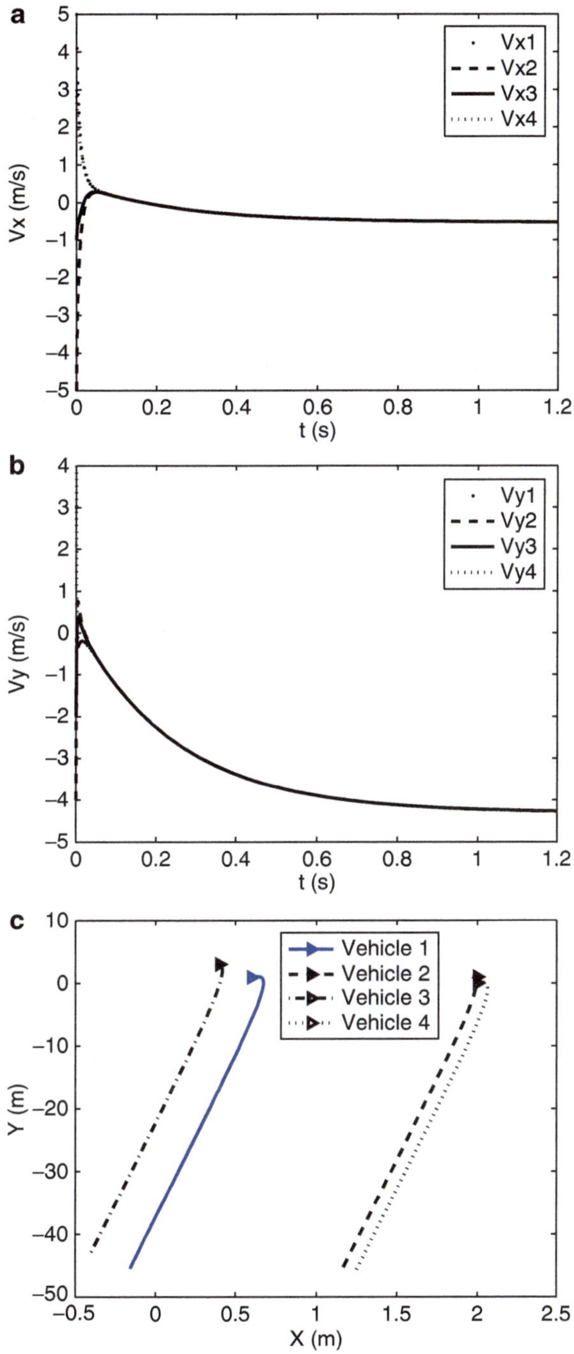

Fig. 3.6 **(a)** The x-component of the velocity profile, **(b)** the y-component of the velocity profile, and **(c)** the $x-y$ path trajectories of the *LL* team of four agents with linear dynamical model resulting from the optimal control strategy in an infinite horizon scenario

solving the set of *LMI*s in (3.49) as $R^1 = 10^{-5}\left(\begin{smallmatrix} 3.28 & -0.027 \\ -0.027 & 5.13 \end{smallmatrix}\right)$, $K^1 = \left(\begin{smallmatrix} 2.013 & -0.39 \\ -0.39 & 0.323 \end{smallmatrix}\right)$, $\Gamma = 10^4\left(\begin{smallmatrix} 1.1879 & 0 \\ 0 & 1.1987 \end{smallmatrix}\right)$.

For the *LL* structure, simulations are conducted under the same configuration as above; however, the objective is to ensure that the team members have the same velocity which is not predefined, but should be in the null space of the matrix $(A^i)^T K^i$. In Fig. 3.6a, b, the x- and y-components of v^i are shown for $i = 1,\ldots,4$. Figure 3.6c depicts the actual path trajectories that are generated by the vehicles in the $x-y$ plane. The initial state of the vehicles and the other simulation parameters are all the same as in the *MLF* case. Also, the parameters corresponding to the model are chosen the same as in the *MLF* structure except for the A^i matrix which is selected as $\left(\begin{smallmatrix} 2 & 4 \\ 1 & 2 \end{smallmatrix}\right)$. The controller gain $K^i = \left(\begin{smallmatrix} 4.45 & -0.79 \\ -0.79 & 0.59 \end{smallmatrix}\right)$ is obtained from solving the Riccati equation for the above configuration. From the simulation results, it may be clearly concluded that the vehicles are aligned and move together with the same desired velocity $v^c = [-0.5 \quad -4.3]^T \in \text{Null}((A^i)^T K^i)$.

3.5 Conclusions

The problem of cooperation in a network of multi-agents with the requirement of achieving consensus was considered for both the *LL* and the *MLF* structures. A semi-decentralized optimal control strategy was designed for a team of multi-agent systems by using minimization of the agents' individual performance cost functions subject to partial availability of local information. An unexpected and interesting outcome of the proposed design strategy is that in an infinite horizon scenario, the control law results in either the well-known average consensus protocol strategy or a modified version of it [79] for both the *LL* and *MLF* structures. In other words, a performance index is introduced that is minimized by the average consensus protocol through the proposed methodology. Corresponding to the *MLF* structure it was assumed that only the leader is aware of the desired command requirements and specifications and that there are corrective feedbacks from the followers to the leader when agents are connected through a prespecified topology.

One of the contributions of the present work compared to the synthesis methods that are introduced in [6, 24, 128] is in the introduction of interaction terms in the dynamical model of the agents to describe and characterize the interconnections and information exchanges among the agents. This novel modelling approach provides a framework in which the local and global control requirements may be partially decoupled. Another advantage of the proposed methodology is robustness of the team to uncertainties and faults in the leader or followers and adaptability of the team members to these unanticipated situations as will be discussed in detail in Chap. 4. Moreover, while optimality of the solutions is guaranteed, given that the optimal control is a multi-objective framework, the proposed method has the added potential advantage of being capable of accommodating other additional specifications, e.g., new timing constraints or limited control input availability.

Chapter 4
Nonideal Considerations for Semi-decentralized Optimal Team Cooperation

In practice, many impediments may prevent the team members to cooperate effectively and efficiently. To address this issue, in this chapter, two nonideal considerations are investigated for the team cooperation problem. Specifically, we will generalize the results that are obtained in the previous chapter to more challenging environments and considerations. We have considered two scenarios. First, the performance of the previously designed team in the presence of actuator faults is investigated. In the second part of this chapter, the control design is modified to address stability and consensus seeking in a switching network topology.

4.1 Team Behavior in Presence of Actuator Faults

In practice, it is quite possible that some agents in a team may become unable to follow the team command due to anomalies and faults. Some sources of this problem can be due to actuator faults or saturations, faults in the measurement of neighbor states, or communication link failures. Due to these malfunctions, the faulty agent cannot follow the command that is provided by the team to achieve the predefined goal. This may result in permanent separation of that agent from the team which correspondingly may affect the cohesion of the team. In this section, we provide results on performance analysis of a team of multi-agent systems subject to actuator faults. The team goal is to accomplish a cohesive motion using the semi-decentralized optimal control that was proposed in Chap. 3. In the following we will investigate the performance of a team with an *MLF* structure subject to three types of faults introduced in Chap. 2. Any fault occurring in the *LL* structure is similar to the leader failure case in the *MLF* structure.

A preliminary version of the material presented below has been published in [104, 107, 109, 115].

E. Semsar-Kazerooni and K. Khorasani, *Team Cooperation in a Network of Multi-Vehicle Unmanned Systems: Synthesis of Consensus Algorithms*, DOI 10.1007/978-1-4614-5073-3_4, © Springer Science+Business Media New York 2013

4.1.1 Team Behavior Subject to the Loss of Effectiveness Fault in an Agent's Actuator

For the team of multi-agent systems that is described in Chap. 2, and for the dynamical equation (2.20) in an *MLF* structure, assume that some of the agents, either some of the followers or the leader, fail to produce the team control command as described in Theorem 3.2. Specifically, due to an actuator *loss-of-effectiveness (LOE)* fault, we now instead have $u^i_f = \alpha u^i, \quad 0 < \alpha \leq 1$, where u^i_f denotes the actual control effort that is applied by the actuator with u^i representing the designed control input. Let us denote the set of failed agents by $A_f = \{i = N-q+1, \ldots, N\}$, and without loss of generality assume that these are the last q agents of the team. If this is not the case, the agents' labels can be easily changed for this purpose. The concatenated velocity vector of the failed agents can be defined as $v_f = [(v^{N-q+1})^T, \ldots, (v^N)^T]^T$. Using the notion of velocity error as introduced in Theorem 3.2, let us define the total error vector corresponding to the healthy agents as e_w and the one corresponding to the faulty agents as e_f.

Now, assume that the closed-loop dynamics of the entire team is described by $\dot{e} = L_{cl}e$, where $e = [(e^1)^T \ \ldots \ (e^N)^T]^T = [e_w^T \ e_f^T]^T$ and

$$L_{cl} = \begin{bmatrix} \Gamma^1 + \beta^1 & \frac{l_{12}}{|N^1|}\Gamma^1 & \cdots & & \frac{l_{1N}}{|N^1|}\Gamma^1 \\ \frac{l_{21}}{|N^2|}\Gamma^2 & \Gamma^2 & \cdots & & \frac{l_{2N}}{|N^2|}\Gamma^2 \\ \vdots & \vdots & \vdots & & \vdots \\ \frac{l_{i1}}{|N^i|}\Gamma^i & \cdots & \frac{l_{ij}}{|N^i|}\Gamma^i & & \cdots \\ \vdots & \vdots & \vdots & & \vdots \\ \frac{l_{N1}}{|N^N|}\Gamma^N & \cdots & \frac{l_{N(N-1)}}{|N^N|}\Gamma^N & \Gamma^N & \end{bmatrix}, \tag{4.1}$$

where l_{ij} is the ijth element of the Laplacian matrix L, and Γ^i and β^1 are defined in (3.33). Let us partition L_{cl} in order to separate the dynamics of the faulty and the healthy agents as follows

$$L_{cl} = \begin{bmatrix} (L_{11})_{m(N-q)\times m(N-q)} & \vdots & (L_{12})_{m(N-q)\times mq} \\ \cdots\cdots & & \cdots\cdots \\ (L_{21})_{mq\times m(N-q)} & \vdots & (L_{22})_{mq\times mq} \end{bmatrix},$$

where m denotes the dimension of v^i (assumed here to be two). Due to the presence of faults, the closed-loop error dynamics will now be changed into

$$\dot{e} = \begin{bmatrix} L_{11} & \vdots & L_{12} \\ \cdots & & \cdots \\ \alpha L_{21} & \vdots & \alpha L_{22} \end{bmatrix} \begin{bmatrix} e_w \\ e_f \end{bmatrix} = L_f e. \tag{4.2}$$

The following lemma shows that the error dynamics will remain stable despite the presence of agents' faults. Moreover, the consensus achievement goal can still be guaranteed and maintained.

Lemma 4.1. *(a)* ***Stability Analysis:*** *For the team of multi-agent systems that is described in Theorem 3.2, when certain agents fail to comply with the designed team control command and instead implement* $u_f^i = \alpha u^i$ *corresponding to the* LOE *fault, the closed-loop error dynamics still remains stable.*

(b) ***Consensus Achievement:*** *Moreover, the velocity error, i.e.,* $e^i = v^i - v^d$, $\forall i$ *will asymptotically approach to zero, and consequently the consensus will still be achieved.*

Proof. The details are provided in Appendix A. □

It is worth noting that although this fault does not deteriorate the stability property of the closed-loop dynamics, it affects the transient behavior of the agents. Specifically, the transient convergence rate becomes dependent on the parameter α.

4.1.2 Team Behavior Subject to an Actuator Float Fault in an Agent

In this section, we analyze the performance of a team of multi-agent systems in the presence of actuator float faults for both the follower agents and the leader agent.

4.1.2.1 Fault Occurrence in the Followers

For a team that is governed by a double-integrator dynamical model of each agent, as described in Theorem 3.2, assume that a number of follower agents fail to produce the team control command as specified by (3.31) and instead one now has $u^i = 0$. The concatenated velocity vector of the failed agents, v_f, is a constant vector (due to u^i being zero).

Now, assume that the closed-loop dynamics of the entire team is described by $\dot{e} = L_{cl}e$, where $e = [e_w^T \ e_f^T]^T$ and L_{cl} are defined as in the previous cases and e_f will be a constant vector if v^d is time-invariant. Partition L_{cl} in order to separate the dynamics associated with the faulty and the healthy agents. Due to the presence of faults, this dynamics is now governed by

$$\dot{e} = \begin{bmatrix} L_{11} & \vdots \ L_{12} \\ \cdots\cdots\cdots & \cdots\cdots \\ 0_{mq\times m(N-q)} & \vdots \ 0_{mq\times mq} \end{bmatrix} \begin{bmatrix} e_w \\ e_f \end{bmatrix}. \tag{4.3}$$

The following lemma demonstrates that the overall team error dynamics will remain stable despite the presence of the followers faults.

Lemma 4.2. *(a)* ***Stability Analysis:*** *For a team of multi-agent systems that is described in Theorem 3.2, if a number of follower agents fail to comply with the team control command as specified by (3.31) and instead implement $u^i = 0$ (corresponding to a float fault), then the closed-loop error dynamics still remains stable and the velocity error, i.e., $e^i = v^i - v^d$, $\forall i$ will remain bounded.*

(b) ***Steady State Error:*** *Moreover, the steady state of the velocity error $e = \begin{bmatrix} e_w \\ e_f \end{bmatrix}$ is $e_{ss} = \begin{bmatrix} -L_{11}^{-1}L_{12} \\ \cdots\cdots \\ I_{mq\times mq} \end{bmatrix} e^f$, where $e^f = v^f - ([1 \; \dots \; 1]_{1\times q}^T \otimes v^d)$ and e^f, v^f are the error and the velocity vectors of the failed agents at the point when their state is frozen.*

Proof. The details are provided in Appendix A. □

4.1.2.2 Fault Occurrence in the Leader Agent

Similar to the discussion presented in the previous section, if the leader has a fault, e.g. its velocity is frozen at a constant value, namely v_f, we can still achieve stability of the team overall error dynamics. The closed-loop system matrix can be partitioned as before. In the presence of the leader fault the closed-loop error dynamics becomes

$$\dot{e} = \begin{bmatrix} 0_{m\times m} & \vdots & 0_{m\times(N-1)m} \\ \cdots & \cdots & \cdots \\ L_{21} & \vdots & L_{22} \end{bmatrix} \begin{bmatrix} e_f \\ e_w \end{bmatrix}. \tag{4.4}$$

The following lemma shows that the error dynamics still remains stable even if a fault has occurred in the leader agent.

Lemma 4.3. *(a)* ***Stability Analysis:*** *For a team of multi-agent systems that is described in Theorem 3.2, if the leader fails to comply with the team control command as specified by (3.32) and instead a zero control is implemented (corresponding to a float fault), i.e., $u^1 = 0$, then the overall team error dynamics still remains stable and the tracking error, i.e., $e^i = v^i - v^d$, $\forall i$ remains bounded.*

(b) ***Steady State Error:*** *Moreover, the final value of the tracking error vector e_{ss} is governed by*

$$e \to e_{ss}, \text{ as } t \to \infty, \; e_{ss} = \mathbf{1} \otimes e^f, \; e^f = v^f - v^d, \tag{4.5}$$

where e^f, v^f are the error and velocity vectors of the leader, respectively, at which its state is frozen and **1** *is the vector of ones.*

Proof. The details are provided in Appendix A. □

Up to this point, it is shown that the team will remain stable in presence of faults. In the conventional leader–follower structure, if a follower fails to follow the leader's command it would be separated from the team and therefore could be lost from the team forever. However, in the *MLF* structure considered here, due to existence of a feedback from the followers to the leader as well as connections among the followers, if one of the followers cannot follow the command, all the other followers and the leader will adapt themselves to this change until this agent is recovered. In this manner the cohesion of the team will be preserved and no member will be lost without affecting the others. In the following, we will show that the healthy agents will adapt themselves to the state of the faulty agent.

4.1.2.3 Leader and Follower Adaptability to Occurrence of Faults

In order to show the network adaptability property under faults we need the following theorem from the graph theory [147]. The theorem provides a relationship corresponding to the second-order minors of the Laplacian matrix of a treelike graph.

Theorem 4.1 ([147]). *Assume that L is the Laplacian matrix of an undirected graph. Denote the minors of the matrix L that is obtained by eliminating the m rows and columns by $L(W|U)$, where W and U are the sets of eliminated rows and columns, respectively. Consider a tree T with N vertices and denote the path [147] between two nodes u and v by $P(u,v)$. Then, for $1 \leq i < j \leq N$ and $1 \leq k < l \leq N$, one has*

$$(-1)^{i+j+k+l}\det(L(i,j|k,l)) = \pm\text{length}(P(v_i,v_j) \cap P(v_k,v_l)). \tag{4.6}$$

Note that the sign of the determinant depends on the relative orientation of $P(v_i,v_j)$ and $P(v_k,v_l)$ with respect to one another in the following sense. If we orient $P(v_i,v_j)$ from v_i towards v_j, and $P(v_k,v_l)$ from v_k towards v_l, they both induce an orientation on their intersection. If these orientations agree, the sign is $+1$; otherwise, it is -1.

Proof. The details are provided in [147]. □

For the sake of proof of the next theorem we now assume that the information structure is described by a treelike graph.

Theorem 4.2 (Leader and Follower Adaptability). *For a team of multi-agent systems that is described in Theorem 3.2, if a follower fails to produce the team control command as specified in (3.31) due to a float fault, i.e., $u^i = 0$, then all the agents will adapt themselves to this agent's change, i.e., the direction of change* *in the state of the faulty follower will be the same as the change in the rest of the team. This implies that the steady state error of the faulty follower and the healthy members will have the same sign, i.e.,*

$$e^k \bullet e^f > 0,\ e^f = v^f - v^d,\ e^k = v^k - v^d,\ k = 1,\ldots,N-1, \tag{4.7}$$

in which v^f denotes the velocity at which the faulty agent's velocity is frozen and e^f denotes the corresponding error, e^k denotes the velocity error of the agent k, "$\bullet$" denotes the Hadamard product [41], and "> 0" refers to positiveness of the vector's elements.

Proof. The proof is provided in Appendix A. □

4.1.3 *Team Behavior Subject to a Lock-in-Place Fault in an Agent*

For a team of multi-agent systems that is described in Theorem 3.2, assume that one of the followers fails to produce the team control command due to a *lock-in-place (LIP)* fault and the applied control input is now frozen at a constant value, i.e., $u_f^i = u_c$, where u_c is a constant value. In this situation, and similar to the discussion provided in the previous section, the closed-loop error dynamics can be represented by the following:

$$\dot{e} = \begin{bmatrix} \dot{e}_w \\ \dot{e}_f \end{bmatrix} = \begin{bmatrix} (L_{11})_{m(N-1)\times m(N-1)} & \vdots & (L_{12})_{m(N-1)\times m} \\ \cdots\cdots & \cdots & \cdots\cdots \\ 0_{m\times m(N-1)} & \vdots & 0_{m\times m} \end{bmatrix} e + \begin{bmatrix} 0 \\ u_c \end{bmatrix}. \tag{4.8}$$

From the above equation it can be concluded that e_f can grow without a bound, and therefore the error dynamics (4.8) is not stable. In other words, in this situation, and given the fact that the open-loop matrix is not asymptotically stable, i.e., $A^i = 0$, one cannot guarantee stability. However, if at least the open-loop matrix of the faulty agent is stable, there might be a possibility to guarantee stability of the error dynamics. Hence, in this section, we first obtain the agent trajectories that are described by the dynamics (2.20). Next, we will analyze the team behavior assuming that the agents have a stable open-loop system matrix. We will discuss both types of agents' dynamical representation as given in Chap. 2, i.e., linear and double integrator models. It is shown that when the open-loop system matrix is asymptotically stable, the stability of the error dynamics is also guaranteed; however, as in the float fault type consensus can no longer be achieved. The following two lemmas summarize our results where an ultimate value of the velocity error vector e_{ss} is also obtained.

Lemma 4.4. *For a team of multi-agent systems that is described in Theorem 3.2, assume that a follower agent fails to comply with the team control command as specified in (3.31) and instead implements $u_f^i = u_c$ (LIP fault), where u_c is a constant vector. The ultimate time-varying value of the velocity error vector e_{ss} can be specified according to*

$$e \to e_{ss} \text{ as } t \to \infty,\ e_{ss} = \begin{bmatrix} -L_{11}^{-1}[L_{12}u_c t - L_{12}u_c t_f + L_{12}e^f + L_{11}^{-1}L_{12}u_c] \\ \cdots\cdots\cdots\cdots\cdots\cdots \\ u_c t - u_c t_f + e^f \end{bmatrix}, \tag{4.9}$$

where t_f denotes the time of fault injection and e^f is the faulty agent velocity error vector at t_f.

Proof. The proof is provided in Appendix A. □

Based on the definition of the float fault, this fault may be considered as a *LIP* fault when $u_c = 0$. Clearly, the steady state error provided in (4.9) is the same as the one given in Lemma 4.2 if u_c is replaced by zero.

For the remainder of our discussions, let us consider a team of multi-agent systems with the governing linear dynamics (2.21) and assume that the faulty agent's open-loop matrix A^f is asymptotically stable. The error dynamics for the entire team can be written as $\dot{e} = L_{cl}e + f(v^d, g^1)$, where L_{cl} is defined as

$$L_{cl} = -K^{-1}(2L \otimes Q^{ij} + A^T K + 2G) = -K^{-1}\Upsilon \tag{4.10}$$

and K, A, G, Υ are defined as in Theorem 3.3. The term $f(v^d, g^1)$ is defined as follows, where it is decomposed corresponding to the healthy and faulty agents' dynamics as

$$f(v^d, g^1) = \begin{bmatrix} f_1(v^d, g^1) \\ \cdots\cdots \\ f_2(v^d, g^1) \end{bmatrix} = \begin{bmatrix} -(K^1)^{-1}(A^1)^T(g^1 + K^1 v^d) \\ -(K^2)^{-1}(A^2)^T K^2 v^d \\ \vdots \\ -(K^i)^{-1}(A^i)^T K^i v^d \\ \vdots \\ \cdots\cdots\cdots\cdots\cdots \\ -(K^f)^{-1}(A^f)^T K^f v^d \end{bmatrix}, \tag{4.11}$$

where superscript f denotes the quantities that correspond to the faulty agent. The rest of parameters are defined as in Chap. 3. When a *LIP* fault occurs in a follower, the closed-loop error dynamics is then governed by

$$\dot{e} = \begin{bmatrix} \dot{e}_w \\ \dot{e}_f \end{bmatrix} = \begin{bmatrix} (L_{11})_{m(N-1)\times m(N-1)} & \vdots & (L_{12})_{m(N-1)\times m} \\ \cdots\cdots\cdots & & \cdots\cdots\cdots \\ 0_{m\times m(N-1)} & \vdots & A^f \end{bmatrix} e + \begin{bmatrix} f_1(v^d, g^1) \\ A^f v^d + B^f u_c \end{bmatrix}, \tag{4.12}$$

where L_{ij} denotes the corresponding partitioning of the matrix $-K^{-1}(2L\otimes Q^{ij}+A^TK+2G)$. In the following lemma, the results on stability and consensus seeking for the above selection of the agents' dynamics are provided.

Lemma 4.5. *(a)* ***Stability Analysis:*** *Consider a team of multi-agent systems with the governing dynamics (2.21) and the control laws for the followers and the leader given by (3.46) and (3.47), respectively. When a follower fails to comply with the designed team control command and instead implements $u^i_f = u_c$, where u_c is a constant value, the closed-loop error dynamics still remains stable if the corresponding open-loop matrix of the faulty agent, i.e., A^f, is Hurwitz.*

(b) ***Steady State Error:*** *Moreover, the ultimate value of the velocity error vector e_{ss} is governed by*

$$e \to e_{ss} \text{ as } t\to\infty,\; e_{ss} = \begin{bmatrix} L_{11}^{-1}[L_{12}((A^f)^{-1}B^f u_c + v^d) - f_1(v^d, g^1)] \\ \cdots\cdots\cdots\cdots\cdots\cdots\cdots\cdots \\ -(A^f)^{-1}B^f u_c - v^d \end{bmatrix}, \tag{4.13}$$

where $f_1(v^d, g^1)$ is defined as

$$f_1(v^d, g^1) = \begin{bmatrix} -(K^1)^{-1}(A^1)^T(g^1 + K^1 v^d) \\ -(K^2)^{-1}(A^2)^T K^2 v^d \\ \vdots \\ -(K^i)^{-1}(A^i)^T K^i v^d \\ \vdots \end{bmatrix} \tag{4.14}$$

and the superscript f denotes the quantities that correspond to the faulty agent.

Proof. The proof is provided in Appendix A. □

4.1.4 Leaderless Structure

The results corresponding to the situation in which one of the members of a leaderless team fails are quite similar to the leader failure in the *MLF* architecture as discussed in the previous sections, and therefore the associated results are not included.

4.2 Switching Network Structure

In many situations two agents in a team may not be able to obtain the state of each other, either through communication links or by means of onboard sensor measurements. This may arise due to either restrictions on their communications, e.g., due to large separating distances or appearance of obstacles among the team members, or it can be due to the changes that are preplanned in the mission of the team. Consequently, according to a specific mission planning and scheduling for a given team, the communication network configuration among the team members may no longer be fixed and therefore could be characterized as a switching network architecture. In this situation, team members have to determine new neighbors in order to maintain the connectivity of the team information graph. This implies that the neighboring sets should be defined as time-varying sets, namely $N^i(t)$. These neighboring sets will result in a set of information graphs with time-varying Laplacian matrices, for which the only assumed condition is their connectivity.

In addition to the changes that may occur in the communication structure of a network, in some circumstances, as in, e.g., the leader–follower structure, the assignment of the leader may also change during the mission. This can arise either as a result of the fact that some agents are more accessible in certain stages of the mission or that for safety considerations some agents are more reliable or safer to be assigned as the leader during certain parts of the mission. Under these conditions, the leader assignment can be time-varying as well. Therefore, the team structure will no longer remain fixed, and consequently one has to analyze the team behavior subject to a switching topology.

Below a *design-based* strategy is proposed which can guarantee consensus achievement for a team of multi-agent systems with a general underlying network graph subject to both network topology as well as the leader assignment changes. In contrast to earlier work in the literature that have focused on analysis of the consensus algorithm subject to a time-varying structure, the approach pursued and presented below is mainly based on utilization of control techniques to design a switching strategy. The proposed framework can handle strongly connected, directed, and unbalanced graphs under a switching network configuration. By assigning the eigenvectors of the closed-loop matrix, which corresponds to the error dynamics of the team to a desirable vector, the existence of a common Lyapunov function and consequently the overall team stability and consensus achievement are all guaranteed simultaneously.

The design strategy for the original fixed team structure is based on the semi-decentralized optimal control approach as defined in Chap. 3. However, with the modifications that are made in the present section, one requires that the control gain matrices that are defined in the cost functions take on specific values. Consequently, our results lead to a constraint on the optimal control law which is designed initially for the fixed network topology. It is shown that by introducing additional criteria the desirable performance specifications of the team can still be ensured and guaranteed. As a demonstration and representation of such a criterion, a performance–control effort trade-off is considered and analyzed in detail.

The preliminary version of the material that is presented in the following sections has already been published in [108, 116].

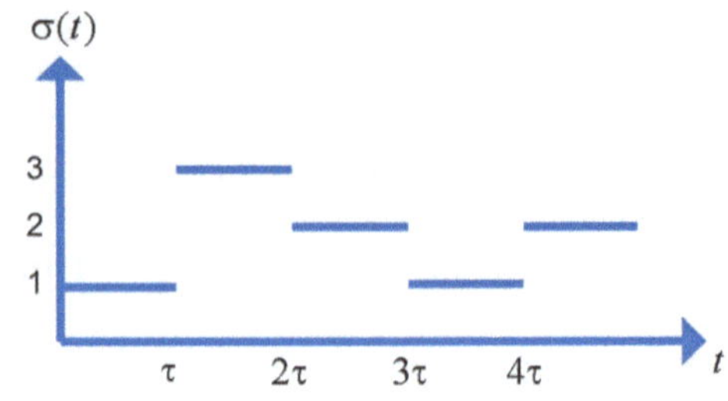

Fig. 4.1 Switching signal $\sigma(t)$

4.2.1 Switching Control Input and Stability Analysis

In this section, we only discuss the *MLF* team structure. The *LL* structure can be treated similarly. Now, assume that the agent's dynamical model is given as in (2.20) and define the error for each agent as $e^i = v^i - v^d$, where the desired leader command v^d is considered to be time-invariant. The error dynamics for the entire team can be obtained as $\dot{e} = L_{cl}e$, where L_{cl} is defined in (4.1). This matrix can be further simplified as follows:

$$L_{cl} = -2K^{-1}\left(L\otimes Q + \begin{bmatrix} \Gamma & 0 & \dots & 0 \\ 0 & \dots & 0 & 0 \\ \vdots & \vdots & \vdots & \vdots \\ 0 & \dots & 0 & 0 \end{bmatrix}\right) = -2K^{-1}(L\otimes Q + G), \tag{4.15}$$

where $K = \mathrm{Diag}\{K^i, i = 1,\dots,N\}$ and $G = \mathrm{Diag}\{\Gamma, 0,\dots,0\}$. The above expression of L_{cl} will be used in the following discussion on stability analysis of the switching topologies.

Now, assume that there is a team of agents that is characterized by a switching topology due to the time-varying neighboring sets $N^i(t)$ or time-varying leader assignment. Associated with this scenario a switching signal that is denoted by $\sigma(t) : \mathbb{R}^+ \to \mathbb{N}$ is defined which is a train of rectangular pulsed signals that has a constant integer value over each time interval τ as shown in Fig. 4.1. The communication links among the agents are assumed to be directional with a Laplacian matrix that is denoted by L. For the case of switching networks, this matrix is a function of the switching signal $\sigma(t)$ and can be written as $L_{\sigma(t)}$, where

$$L_\sigma \in \{L | L \text{ describes the Laplacian of a strongly connected digraph}\}. \tag{4.16}$$

Hence, during each time interval, the Laplacian matrix describing the underlying team architecture graph belongs to the family of Laplacian matrices that is defined in (4.16).

Assumption 4.1 *The Laplacian matrix $L_{\sigma(t)}$ is provided and available to all the agents of the network.*

This assumption will be further used to evaluate the switching control signal. It is worth noting that providing this information to the individual agents does not impose an extra restriction on the semi-decentralized structure of our proposed control strategy. In fact regardless of this assumption, to ensure and verify the connectedness requirement of the information exchange graph, each agent should already switch its communication links so that the entire network remains connected. Therefore, each agent should be aware of both the local as well as the global connections. In other words, the requirement that each agent is aware of the Laplacian of the team is not an impediment for or an extra restriction on our proposed semi-decentralized control strategy.

To emphasize that the leader assignment is time-varying, i.e., agent 1 is not necessarily the leader, we may assume that the matrix G in (4.15) is a function of the switching signal $\sigma(t)$ as well, i.e., G_σ. Depending on the agent that is assigned as the leader, the corresponding row in the matrix G will be nonzero. Subsequently, we denote the parameters that are associated with a switching by the subscript $\sigma(t)$, i.e., $(.)_\sigma$. Therefore, the closed-loop matrix defined in (4.15) is rewritten as

$$L_{cl,\sigma} = -2K_\sigma^{-1}(L_\sigma \otimes Q + G_\sigma),$$

where $L_{cl,\sigma}, K_\sigma, L_\sigma, G_\sigma$ denote the matrices L_{cl}, K, L, G corresponding to the switching structure, respectively. Obviously, the controller coefficient matrix K depends on the switching state since K is a function of the neighboring sets $N^i(t)$. If $L_{cl,\sigma}$ was describing the Laplacian of a balanced graph, then it was straightforward to find a common Lyapunov function for the entire switching network and correspondingly prove the stability of the overall network. However, in the present case, this matrix does not have the above-mentioned property. In the following, we design a switching controller such that the network stability can be guaranteed and maintained.

Towards this end, let us partition the matrix $L_{cl,\sigma}$ into two parts, namely $\bar{L}_\sigma = K_\sigma^{-1}(L_\sigma \otimes Q)$ and $K_\sigma^{-1}G_\sigma$. The first part, $\bar{L}_\sigma$, is itself the Laplacian of a directed weighted graph which is not necessarily balanced. However, if we could transform $\bar{L}_\sigma$ into the Laplacian of a balanced graph, then it is easy to show that a common Lyapunov function for the corresponding switching system exists. One solution to achieve the above goal is to design a *switching control* law such that $\bar{L}_\sigma$ becomes the Laplacian of a balanced graph for any switching network. This implies that we need to modify the design of the matrix K_σ, as given in Lemma 3.2 and Theorem 3.2, such that $\bar{L}_\sigma$ satisfies the required property. One way to design K_σ to compensate for the switching structure is by selecting different Q^{ij}s for different nodes in each switching structure (in contrast to the assumption in Theorem 3.2). If such a control design goal can be accomplished, not only the undirected graphs but also the directed and unbalanced graphs can be analyzed under the switching network topology assumption.

Towards this end, let us assume that Q^{ij} is no longer equal to Q and has different values that are denoted by $Q^i_\sigma(t)$ for each agent i and each switching state $\sigma(t)$ so that $\bar{L}_\sigma$ can be written as follows:

$$\bar{L}_\sigma = \begin{bmatrix} (K_\sigma^1)^{-1}Q_\sigma^1 l_{11} & (K_\sigma^1)^{-1}Q_\sigma^1 l_{12} & \dots & (K_\sigma^1)^{-1}Q_\sigma^1 l_{1N} \\ (K_\sigma^2)^{-1}Q_\sigma^2 l_{21} & (K_\sigma^2)^{-1}Q_\sigma^2 l_{22} & \dots & (K_\sigma^2)^{-1}Q_\sigma^2 l_{2N} \\ \vdots & \vdots & \vdots & \vdots \end{bmatrix},$$

where l_{ij} is the ijth entry of the matrix L_σ which is time dependent (due to the assumption of switching topology) and K_σ^i is the matrix K^i corresponding to the switching network structure. In order to have $\bar{L}_\sigma$ as a balanced matrix, we should have

$$(\mathbf{1}^T \otimes I_n)\bar{L}_\sigma = \mathbf{0} \to \left[(K_\sigma^1)^{-1}Q_\sigma^1 l_{11} \; (K_\sigma^2)^{-1}Q_\sigma^2 l_{22} \; \dots \; (K_\sigma^N)^{-1}Q_\sigma^N l_{NN} \right]$$
$$\times \left(\begin{bmatrix} 1 & l_{12}/l_{11} & \dots & l_{1N}/l_{11} \\ l_{21}/l_{22} & 1 & \dots & l_{2N}/l_{22} \\ \vdots & \vdots & \vdots & \vdots \end{bmatrix} \otimes I_n \right) = \mu_\sigma^T(\hat{L}_\sigma \otimes I_n) = 0, \tag{4.17}$$

where n denotes the dimension of the agents' output, $\hat{L}_\sigma$ denotes the normalized Laplacian matrix of the graph, and μ_σ is defined as

$$\mu_\sigma = \left[(K_\sigma^1)^{-1}Q_\sigma^1 l_{11} \; (K_\sigma^2)^{-1}Q_\sigma^2 l_{22} \; \cdots \; (K_\sigma^N)^{-1}Q_\sigma^N l_{NN} \right]^T. \tag{4.18}$$

To ensure that the expression (4.17) is satisfied, Q^{ij} $(Q_\sigma^i), R^i$ (R_σ^i), and Γ (Γ_σ) should be selected such that μ_σ in (4.18) will be in the left null-space of $\hat{L}_\sigma \otimes I_n$. Assume that ω_σ is a normalized vector in the left null-space of $\hat{L}_\sigma$ (the eigenvector of $\hat{L}_\sigma$ corresponding to the zero eigenvalue), then we should have

$$\mu_\sigma = \kappa \omega_\sigma \otimes I_n, \tag{4.19}$$

where κ is a scaling factor that should be selected by using a specific criterion, e.g. along the lines that are provided in the next section. Therefore, (4.19) is the main requirement that should be satisfied by proper selection of Q_σ^i, R_σ^i, and Γ_σ. Now, we state the following lemma which is used in the subsequent discussions.

Lemma 4.6. *The Laplacian matrix of any strongly connected directed graph has a left eigenvector which corresponds to the zero eigenvalue and whose entries have the same sign, i.e., they are either all positive or all negative.*

Proof. The details are provided in Appendix A. □

We are now in a position to summarize the previous discussions into the following theorem.

Theorem 4.3 (Stability Analysis Under Switching Structure). *For a team of mobile robots that is described in Lemma 3.2, and under the assumptions of switching network and switching leader, the control laws $u_\sigma^i, i = 1, \dots, N$ selected according to*

$$u_\sigma^i(v^i, v^j) = \Gamma_\sigma^i \left(v^i - \frac{\sum_{j \in N^i} v^j}{|N^i(t)|} \right), \; i = 2, \ldots, N, \tag{4.20}$$

$$u_\sigma^1(v^1, v^j) = \Gamma_\sigma^1 \left(v^1 - \frac{\sum_{j \in N^1} v^j}{|N^1(t)|} \right) + \beta_\sigma^1 (v^1 - v^d), \tag{4.21}$$

$$\Gamma_\sigma^i = -2\kappa\rho_{i,\sigma} I_n, \; i = 1, \ldots, N, \; \beta_\sigma^1 = -\gamma \left(\kappa\rho_{1,\sigma} r^1 + \sqrt{(\kappa\rho_{1,\sigma} r^1)^2 + \gamma r^1} \right) I_n \tag{4.22}$$

will guarantee that the cost functions in (3.5) and (3.6) are minimized if the parameters Q^{ij} in these cost functions are selected as $Q^{ij} = q^i I$, $\forall i$. Furthermore, q^i is obtained from the following equations:

$$|N^1|^2 (q^1)^2 - (4\kappa^2 \rho_{1,\sigma}^2 |N^1| r^1) q^1 - 4(\kappa\rho_{1,\sigma})^2 \gamma r^1 = 0,$$

$$q^i = \frac{4(\kappa\rho_{i,\sigma})^2 r^i}{|N^i|}, \; i = 2, \ldots, N, \tag{4.23}$$

where $\rho_{i,\sigma}$ is the ith element of the vector ω_σ, i.e., an eigenvector of the normalized Laplacian matrix of the graph $\hat{L}_\sigma$, corresponding to its zero eigenvalue. The matrices R^i and Γ that are used in the cost functions (3.5) and (3.6) are chosen as $R^i = r^i I$, $\Gamma = \gamma I$, where r^i, γ are two positive constants and κ is a design parameter. This in turn guarantees that for the family of the closed-loop error dynamics

$$\dot{e} = L_{cl,\sigma} e, \; L_{cl,\sigma} = -2K_\sigma^{-1} (Q_\sigma L_\sigma \otimes I_n + G_\sigma), \tag{4.24}$$

a common Lyapunov function exists. This function ensures that the closed-loop dynamics is asymptotically stable, where $e = [(e^1)^T \; \ldots \; (e^N)^T]^T$, $e^i = v^i - v^d$, and $Q_\sigma = \operatorname{diag}\{Q_\sigma^1, \ldots, Q_\sigma^N\}$. Therefore, the team consensus is achieved under a switching network topology, i.e., $v^i \to v^d$, $\forall i$.

Proof. Refer to Appendix A for details. □

Remark 4.1. It is worth noting that in the above theorem theoretically there are no constraints on the switching signal as it can be selected arbitrary and with any desired frequency characteristics. However, from practical viewpoint and considerations, the switching frequency should be selected according to the dynamic range of the actuators that are employed for implementing the corresponding switching control laws. In other words, the physical constraints that are imposed by the practical specifications of the actuators should be considered in the selection of the switching signal.

Given that the performance of the optimal controller is now limited due to the additional constraints that are imposed on the cost function gains $Q^{ij}(Q_\sigma^i)$ as in (4.23), one may compensate for this performance degradation by introducing a new criterion for selecting the design parameter κ. This parameter can be considered as a scaling factor which can play the role of defining weights given to various design specifications. Different criteria can be considered in order to guarantee a specific closed-loop behavior. One such criterion deals with a trade-off between the control performance and the control effort, namely the relationship between the matrices $Q^{ij}(Q^i_\sigma)$ and R^i as discussed in the following section.

4.2.2 *Selection Criterion for κ: Performance–Control Effort Trade-Off*

An issue that we would like to consider here deals with defining the criterion for selecting the scaling factor κ. One such criterion may be specified by making a trade-off between the control performance and the control effort. According to the definitions of the cost functions given in (3.5) and (3.6), Q^{ij} defines the weight that is assigned to the performance, whereas R^i is the weight that is assigned to the control effort. Therefore, depending on the specifics of an application, the selected weights can be assigned differently. For example, one may require a predefined ratio between the matrices Q^{ij} and R^i, i.e. one may require that $\frac{\lambda_{\max}(Q^{ij})}{\lambda_{\max}(R^i)} > m_i$, where m_i denotes the desirable value describing the trade-off between the performance and the control effort gain matrices. The following lemma provides sufficient conditions for guaranteeing this requirement.

Lemma 4.7. *In a switching network topology as described in Theorem 4.3, to achieve a trade-off between the performance–control effort in the cost function (3.5) as manifested by* $\frac{\lambda_{\max}(Q^{ij})}{\lambda_{\max}(R^i)} > m_i$, $i = 1, \ldots, N$, *the design parameter* κ *as defined in Theorem 4.3 should be selected according to*

$$\kappa^2 > \frac{1}{4}\max\left\{\frac{m_1|N^1|}{\rho^2_{1,\sigma}}, \frac{\max_{i=2,\ldots,N}(m_i|N^i|)}{\min_{i=2,\ldots,N}(\rho^2_{i,\sigma})}\right\}, \tag{4.25}$$

where m_i *denotes the desirable value describing the trade-off between the performance and the control effort gain matrices.*

Proof. Refer to Appendix A for details. □

4.3 Simulation Results

In this section, the analytical and theoretical results that are obtained in the previous sections are verified through performing simulation studies corresponding to several nonideal scenarios that may occur for a team of multi-agent systems.

4.3.1 *Effects of Actuator Faults on the Team Performance*

In this section, simulation results are presented for the *LOE*, the *LIP*, and the float faults that occur in one of the vehicles in a team of four mobile robots. Without loss of generality, the ring topology is considered for the *MLF* team.

4.3.1.1 LOE Fault

Simulations are conducted for the agents' dynamics as given in (2.20). The leader command is assumed to be a pulse-like signal with a duration of 50 s and its value switches between $v^d = [3\ \ 4]^T$ and $v^d = [5\ \ -1]^T$. The state vector of each agent X^i is composed of the position and the velocity vectors, i.e., $X^i = [(r^i)^T,\ \ (v^i)^T]^T$ and position and velocity vectors are two-dimensional, i.e., $r^i = [r^i_x,\ \ r^i_y]^T$ and $v^i = [v^i_x,\ \ v^i_y]^T$. The initial state of the vehicles are selected as $X^1_0 = [6\ \ 1\ \ 5\ \ 3]^T$, $X^2_0 = [2\ \ 4\ \ -5\ \ -4]^T$, $X^3_0 = [4\ \ 3\ \ -1\ \ -2]^T$, and $X^4_0 = [2\ \ 0\ \ 3\ \ 4]^T$ and the other parameters are selected according to $Q^{ij} = \left[\begin{smallmatrix} 1 & 0 \\ 0 & 3 \end{smallmatrix}\right]$, $R^i = I_{2\times 2}$, $|N^i| = 2$, and $\Gamma = \left[\begin{smallmatrix} 10 & 0 \\ 0 & 40 \end{smallmatrix}\right]$. Figure 4.2a, b show the x- and the y-components of the velocity profiles of the agents when the fourth agent is injected with an *LOE* fault during the period $115 \le t \le 135$. In this period, the fourth agent's actuator is set to $u^4_f = 0.5u^4$. It can be seen that the occurrence of the fault affects the team performance in a short time period and soon thereafter the team recovers its cohesion and achieves consensus. Figure 4.2c shows the actual path trajectories that are generated by the vehicles in the x-y plane. The above results verify the accomplishment of stability and consensus of the team subject to actuator *LOE* faults as obtained in Sect. 4.1.1.

4.3.1.2 Float Fault

The results that are shown in this section are conducted to capture the average behavior of our proposed control strategies through Monte Carlo simulations. The average team response due to 30 different randomly selected initial conditions is presented for the agents' dynamics as given in (2.20). Figures 4.3a, b show the x- and y-components of the average velocities of the agents when a float fault occurs in the third agent during the period $20 \le t \le 30$. In this period, the third agent cannot follow the team command ($v^d = [3\ \ 4]^T$) and its velocity is frozen at $v^3 = [6\ \ 1]^T$. It can be seen that the other agents modify their speed to keep the team cohesion. Figure 4.3c shows the x-y path that is generated in this case. The value to which the team vehicles' average velocity error converges to is $e_{ss} = [1.3\ \ -1.29\ 2.14\ \ -2.15\ 2.14\ \ -2.15\ 3\ \ -3]^T$, which is compatible with the result that is given in Lemma 4.2.

The simulation parameters that are selected are $Q^{ij} = \left[\begin{smallmatrix} 1 & 0 \\ 0 & 3 \end{smallmatrix}\right], R^i = I_{2\times 2}, |N^i| = 2$, $E^i = 0.5I_{2\times 2}$, and $\Gamma = \left[\begin{smallmatrix} 1.3 & 0 \\ 0 & 4 \end{smallmatrix}\right]$, and the random initial conditions for the Monte Carlo simulations are considered as $X^1_0 = [r(0,15)\ \ r(0,25)\ \ r(-5,0)\ \ r(-6,-1)]^T, X^2_0 =$

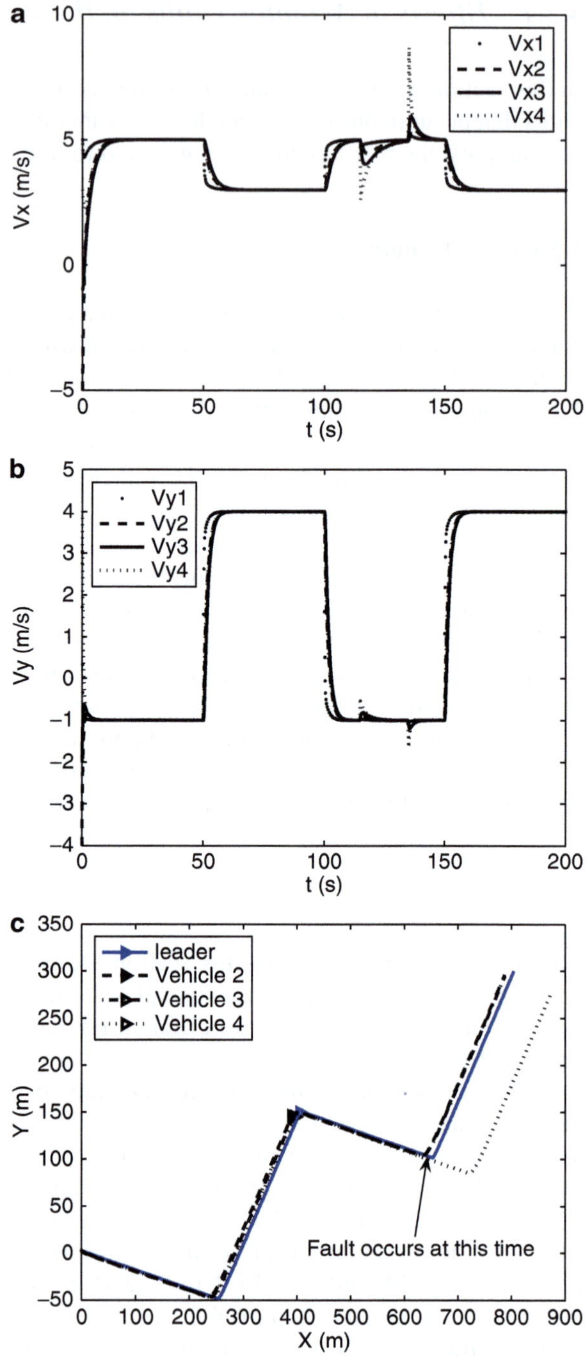

Fig. 4.2 **(a)** The x- and **(b)** the y-components of the velocity profiles and **(c)** the x-y path trajectories of the *MLF* team of four agents in presence of the *LOE* fault in the fourth vehicle during $115 \leq t \leq 135$, where $u_f^4 = 0.5u^4$

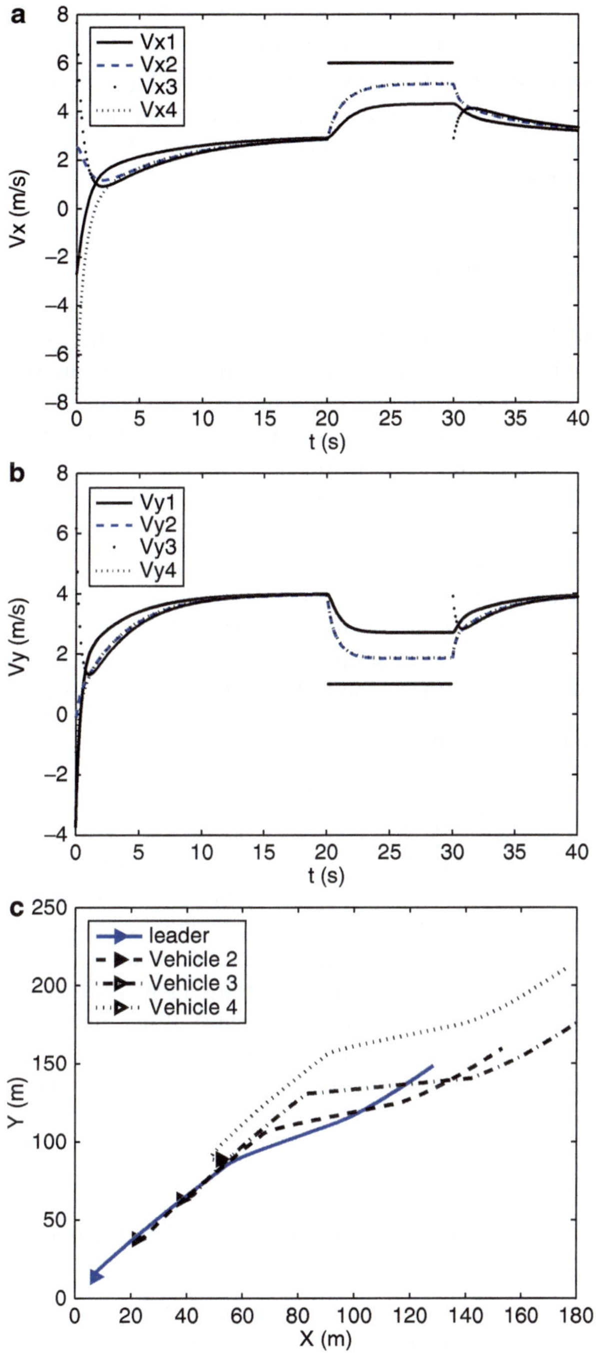

Fig. 4.3 **(a)** The x-component, **(b)** the y-component, and **(c)** the x-y path trajectories of the *MLF* team of four agents (Monte Carlo simulation runs) in the presence of a float fault in the third vehicle (velocity is frozen at $v^3 = [6\ 1]^T$). The jump in the velocity of agent 3 at $t = 30$ s is due to the initiation of a recovery procedure in the actuator of agent 3 (following the fault that is injected at $t = 20$ s) to the healthy velocity after $t \geq 30$ s

$[r(15,30)\ r(25,50)\ r(0,5)\ r(-5,5)]^T, X_0^3 = [r(30,45)\ r(50,75)\ r(5,10)\ r(1,11)]^T$, and $X_0^4 = [r(45,60)\ \ r(75,100)\ \ r(-10,-5)\ r(-6,4)]^T$, where $r(x,y)$ denotes a random variable in the interval $[x,y]$. Figures 4.4a, b depict the average v_x^i, v_y^i trajectories when the velocity of the third agent is kept at $v^3 = [0\ 3]^T$ (the agent does not move in the x direction). Figure 4.4c shows the x-y path that is generated in this case.

4.3.1.3 LIP Fault

In this section, simulations are conducted for the agents' dynamics as given in (2.20) and (2.21). For the former case, all the settings are the same as the ones used in the case of *LOE* fault except for the command pulse duration which is selected as 20 s. Figure 4.5a, b show the x- and y-components of the velocity profiles of the agents when the third agent is injected with the *LIP* fault during the period $20.5 \leq t \leq 25$. During this period, the third agent's actuator is set to $u_f^3 = u^3(t = 20.5)$. It can be seen that after the occurrence of the fault the agents' velocity diverges to different values as predicted by Lemma 4.4. Figure 4.5c shows the actual path trajectories that are generated by the vehicles in the x-y plane.

For the linear dynamical model (2.21), the command and the initial state of the vehicles are similar to the previous case. Other parameters are selected to be $Q^{ij} = 100I_{2\times2}$ and $R^i = 0.01I_{2\times2}$ for $i = 2,\ldots,N$. The parameters corresponding to the model are chosen as $\bar{A}^i = \begin{bmatrix} 1 & 0 \\ 2 & 6 \end{bmatrix}, A^i = -I_{2\times2}$, and $B^i = \begin{bmatrix} 1 & 1 \\ 2 & 6 \end{bmatrix}$. The matrices R^1, K^1, Γ are obtained by solving the set of *LMI*s in (3.49) as $R^1 = 10^5 \begin{bmatrix} 0.10 & 0.17 \\ 0.17 & 1.27 \end{bmatrix}, K^1 = 10^4 \begin{bmatrix} 8.35 & -1.22 \\ -1.22 & 7.09 \end{bmatrix}$, and $\Gamma = 10^5 \begin{bmatrix} 2.35 & 0 \\ 0 & 1.27 \end{bmatrix}$. Figure 4.6a shows the actual path trajectories that are generated by the vehicles in the x-y plane. Figure 4.6b, c show the x- and y-components of the velocity profiles of the agents when the third agent is injected with the *LIP* fault during the period $20.5 \leq t \leq 25$. During this period, the third agent's actuator is set to $u_f^3 = u^3(t = 20.5)$. It can be seen that after the occurrence of the fault the agents' velocities converge to values that are different from the set point but are finite. This verifies the stability and boundedness of the agents' velocities subject to the actuator *LIP* fault as governed by Lemma 4.5.

4.3.2 Team Performance in a Switching Network Topology

Simulation results that are presented in this section are for a team of four agents. The team structure switches between three structures based on a specific switching signal pattern that is shown in Fig. 4.1. It follows from this figure that the switching signal can take three different values at different time intervals, namely 1, 2, and 3.

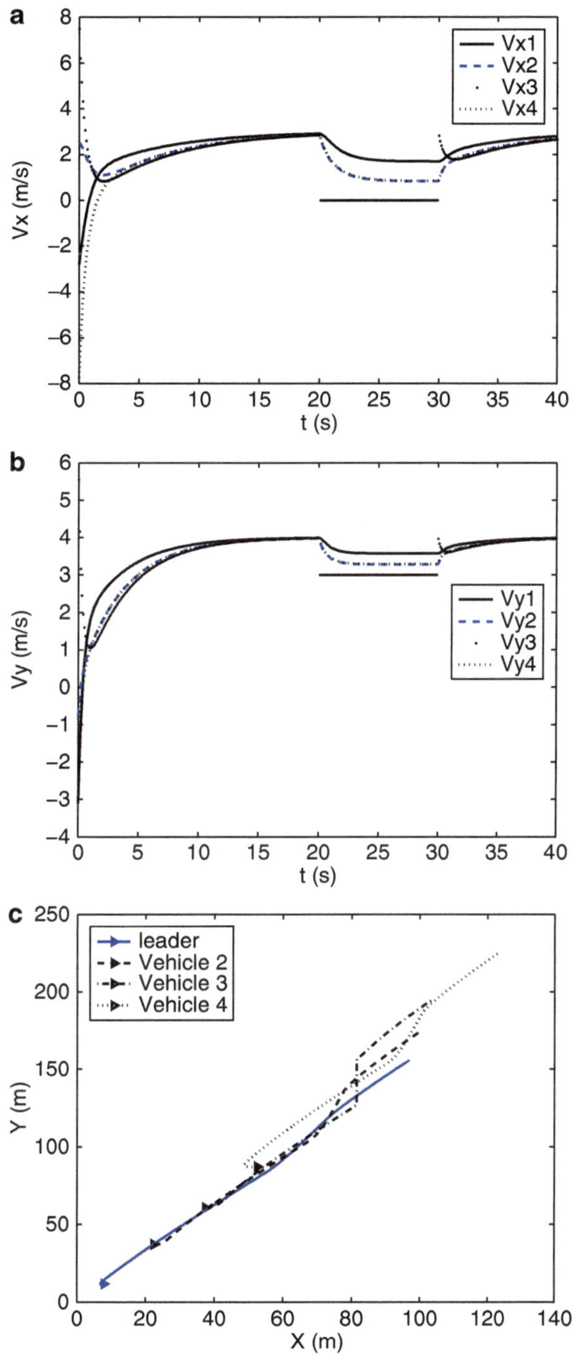

Fig. 4.4 **(a)** The x-component of the average velocity profiles, **(b)** the y-component of the average velocity profiles, and **(c)** the x-y path trajectories of the *MLF* team of four agents (Monte Carlo simulation runs) in presence of the float fault in the third vehicle (velocity is frozen at $v^3 = [0\ 3]^T$ during $20 \leq t \leq 30$)

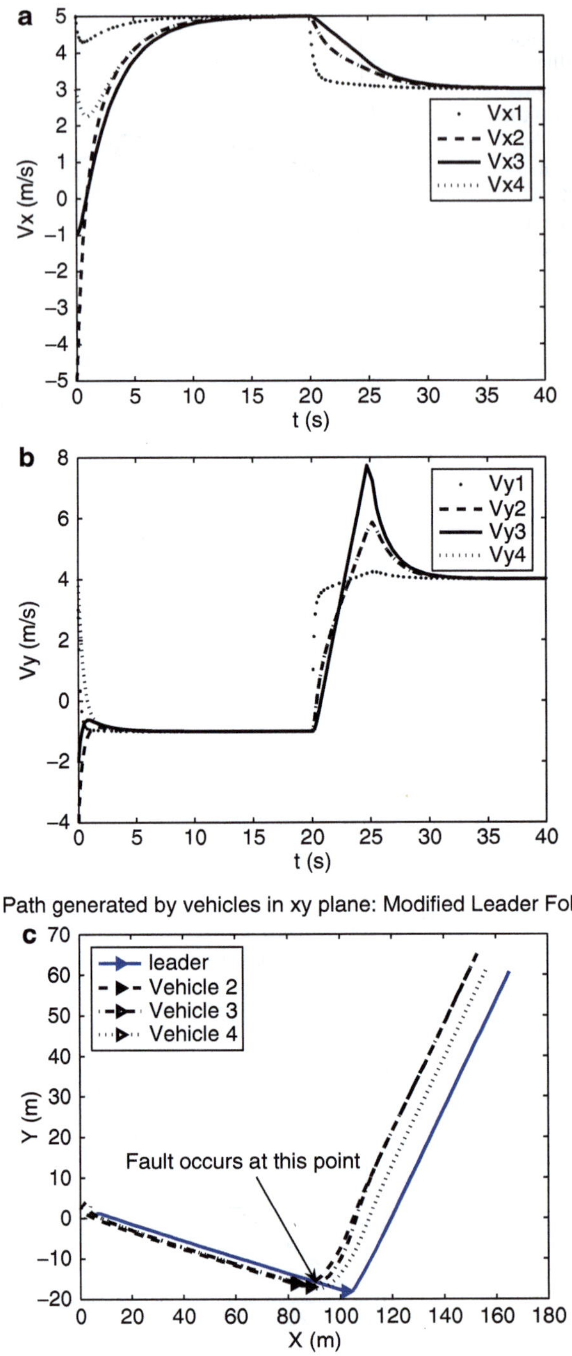

Fig. 4.5 **(a)** The x-component of the velocity profiles, **(b)** the y-component of the velocity profiles, and **(c)** the x-y path trajectories for the *MLF* team of four agents in presence of the *LIP* fault in the third vehicle during $20.5 \leq t \leq 25$ where $u_f^3 = u^3(t = 20.5)$

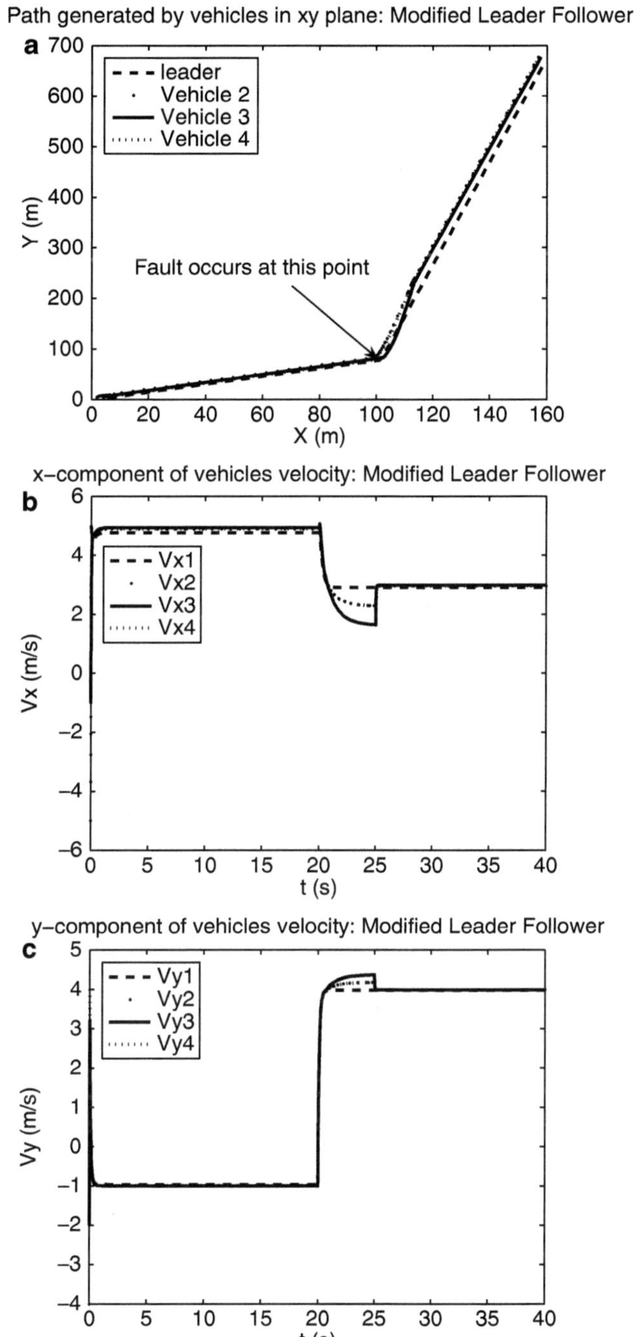

Fig. 4.6 **(a)** The *x*-*y* path trajectories, **(b)** the *x* component of the velocity profiles, and **(c)** the *y* component of the velocity profiles of the *MLF* team of four agents with a linear model in presence of the *LIP* fault in the third vehicle during $20.5 \leq t \leq 25$ where $u_f^3 = u^3(t = 20.5)$

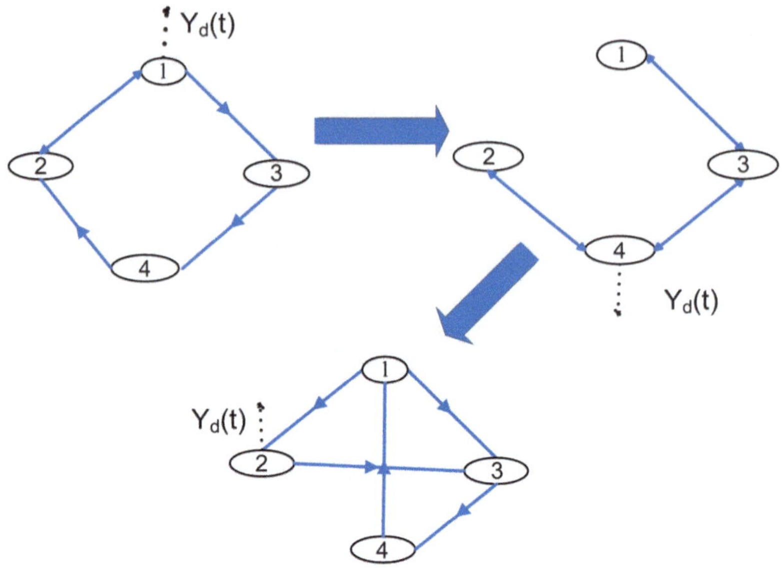

Fig. 4.7 The structure associated with the dynamic transitions of the team for three different switching topologies and leader assignments

In other words, there are three different states for the team structure and the leader assignment during the mission. The leader assignment is changing at each switching instant and is defined to be according to agents 1, 4, and 2 corresponding to $\sigma(t) = 1,2,3$, respectively. Moreover, the leader command is a pulse-like signal which has the same duration as the switching signal time interval, τ. The leader commands for $\sigma(t) = 1,2,3$ is given by $v^d = [15\ 14]^T,\ [7\ 20]^T,\ [20\ 6]^T$, respectively. The graphs describing the network structure are directional and the Laplacian matrices corresponding to the three switching states are as follows:

$$L_1 = \begin{pmatrix} 1 & -1 & 0 & 0 \\ -1 & 2 & 0 & -1 \\ -1 & 0 & 1 & 0 \\ 0 & 0 & -1 & 1 \end{pmatrix}, L_2 = \begin{pmatrix} 1 & 0 & -1 & 0 \\ 0 & 1 & 0 & -1 \\ -1 & 0 & 2 & -1 \\ 0 & -1 & -1 & 2 \end{pmatrix}, L_3 = \begin{pmatrix} 1 & 0 & 0 & -1 \\ -1 & 1 & 0 & 0 \\ -1 & -1 & 2 & 0 \\ 0 & 0 & -1 & 1 \end{pmatrix}. \tag{4.26}$$

The structure associated with the dynamic transitions of the team is shown in Fig. 4.7.

The simulation results are obtained by applying the switching control laws that are given in Theorem 4.3 to the agents with the dynamics that is governed by (2.20). In Fig. 4.8a, the x-component and in Fig. 4.8b, the y-component of the velocity profiles of the four-agent team are shown for the above configurations. Figure 4.8c shows the paths that are generated by the agents during the mission where the team members are switching to different structures and operating under

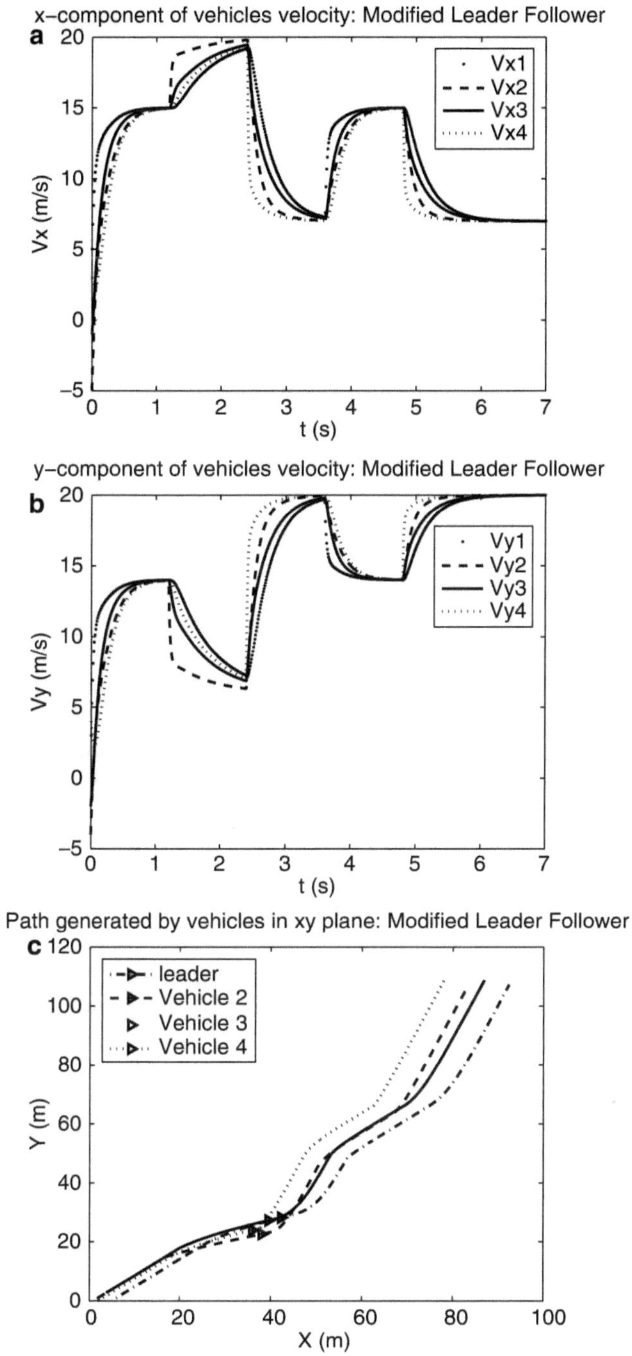

Fig. 4.8 **(a)** The x-component and **(b)** the y-component of the velocity profiles and **(c)** the x-y path trajectories of the *MLF* team of four agents with switching structure and switching leader that are obtained by the application of our proposed switching control strategy

different commands and leaders. It can be seen that the team goal, i.e., consensus achievement, is guaranteed in the presence of the switching topology and switching leader.

4.4 Conclusions

In this chapter, two nonideal considerations associated with the cooperative control of a team of multi-agent systems are analyzed as specifically related to actuator faults and switching network topologies.

First, we provided a formal analysis and an insight into the effects of various actuator faults on the performance of a team of multi-agent systems. It was shown that presence of the *LOE* fault in one of the agents does not deteriorate the stability or the consensus-seeking goal of the team. This fault will only result in a different transient behavior, such as a change in the agent's convergence rate, without a change in the consensus value. On the other hand, if the fault in one or more of the agents is of the float type, either in the leader or the followers, the team does not maintain its consensus objective any longer; however, the stability of the team can still be guaranteed. Moreover, the leader and the healthy followers adapt themselves to the changes when a float fault occurs in one of the agents. In this manner cohesion and cooperation of the team is maintained until the fault is recovered. Finally, the behavior of the team in the presence of a *LIP* fault was also investigated. It was shown that stability of the team can be guaranteed if the agent open-loop system matrix is stable, but the consensus can no longer be achieved. Under the scenario of a *LIP* or a float fault the steady state error is also analytically obtained.

In the second part of this chapter, a semi-decentralized optimal control design strategy for consensus seeking in a team of multi-agent systems with both the switching structure and the dynamic leader assignment was presented. In contrast to the common assumptions in the literature where graphs are assumed to be balanced, here it was assumed that the graph describing the communication topology is not necessarily a balanced graph. A criterion for selecting the controller parameters was proposed to guarantee a specific cooperative control performance requirement.

Chapter 5
Linear Matrix Inequalities in the Team Cooperation Problem

In this chapter, we utilize the linear matrix inequality (*LMI*) formulation to solve the consensus-seeking problem in two frameworks, namely game theoretic and optimal control. In both approaches we take the advantage of the *LMI*'s tool to formulate the presence of relevant constraints on the consensus-seeking problem. In other words, certain constraints are added to the original consensus-seeking problem in order to yield a decentralized solution that guarantees consensus achievement. The first part of this chapter is dedicated to the game theoretic-based approach and the second part discusses the optimal control approach that is based on a novel state decomposition concept which is formally specified subsequently.

5.1 A Cooperative Game Theory Approach to Consensus Seeking

In this section, the methodology for solving the consensus-seeking problem is based on the cooperative game theory. However, in order to clarify the cooperative nature of the method and for comparative purposes, first the semi-decentralized optimal control strategy that is based on minimization of individual costs, as introduced in Chap. 3, is utilized. Cooperative game theory is then used to ensure team cooperation by considering a linear combination of the individual agent costs as a team cost function. The cooperative game theory framework has the advantage of being a multi-objective design tool which is well suited for the problem under consideration in this book. Moreover, this approach guarantees a *cooperative* solution when compared to the other multi-objective design tools.

Minimization of the team cost function results in a set of Pareto-efficient solutions. The choice of the Nash bargaining solution (NBS) among the set of Pareto-efficient solutions guarantees the minimum individual cost. The *NBS* is obtained by maximizing the product of the difference between the costs achieved through the semi-decentralized optimal control strategy and the one obtained

E. Semsar-Kazerooni and K. Khorasani, *Team Cooperation in a Network of Multi-Vehicle Unmanned Systems: Synthesis of Consensus Algorithms*,
DOI 10.1007/978-1-4614-5073-3_5, © Springer Science+Business Media New York 2013

through the Pareto-efficient solution. The latter solution results in a lower cost for each agent at the expense of requiring full information set. To avoid this drawback certain constraints are added to the structure of the controller which is suggested for the entire team by using the *LMI* formulation of the minimization problem. Furthermore, the consensus achievement condition is added as a constraint to the set of *LMI*s. Consequently, although the controller is designed to minimize a unique team cost function, it only uses the available information set associated with each agent. A comparison between the average cost that is obtained by using the above two methods is obtained and analyzed.

5.1.1 Problem Formulation

According to the discussions in Chap. 2, cooperation in a team of N players (agents), as in consensus seeking, can be solved in the framework of cooperative games. The goal in this section is to develop a cooperative solution that utilizes decentralized cost functions and combines them in a team cost function. This will ensure improvements in minimizing individual costs by utilizing the game theoretic method. Towards this end, our objective is to find a set of Pareto optimal solutions for minimizing the team cost function through solving the minimization problem in (2.39). An *NBS* solution can be selected among this set of Pareto-efficient solutions by solving the maximization problem in (2.41) [or (2.40)].

Assume a leaderless *(LL)* team of agents where the dynamical model of each agent and the related cost functions are described in (2.6) and (3.5), respectively. The first step is to combine the individual cost functions in (3.5) into a team cost function J^c according to the following:

$$J^c = \sum_{i=1}^{N} \alpha^i J^i(U) = \int_0^T [Y^T \hat{Q} Y + U^T R U] \mathrm{d}t = \int_0^T [X^T Q X + U^T R U] \mathrm{d}t, \tag{5.1}$$

in which $\alpha = (\alpha^1, \ldots, \alpha^N) \in \mathscr{A}$ as defined in Chap. 2, $J^i(U)$ is the cost function for the ith agent (player) that is defined in (3.5), and $U(\alpha) = [(u_l^1)^T \ \ldots (u_l^N)^T]^T$ is the vector of all the agents' local input vectors. The variables X, Y are the vectors of the entire team state and output as defined in Chap. 2, and other parameters are defined as follows:

$$R = \mathrm{Diag}\{\alpha^1 R^1, \ldots, \alpha^N R^N\}, \ \hat{Q} = [\delta_{hk}]_{N\times N}, \ Q = C^T \hat{Q} C, \tag{5.2}$$

where

$$\delta_{hh} = \sum_{j\in \mathscr{N}^h} \alpha^j Q^{jh} + \alpha^h \sum_{k\in N^h} Q^{hk}, \delta_{hk} = \begin{cases} -\alpha^h Q^{hk} - \alpha^k Q^{kh} & \text{for } k \in N^h, \\ 0 & \text{otherwise,} \end{cases}$$

where N^h is the neighboring set of agent h, $\mathscr{N}^h$ denotes the set of indices of the neighboring sets to which agent h belongs to, and C is defined in (2.8). Each agent belongs to only those clusters in which one of the agent's neighbors exists. Therefore, the total number of these clusters is the same as the number of the neighbors of that agent, i.e., $\mathscr{N}^h = N^h$.

The associated dynamical model constraint of the team is given by (2.7) and (2.8). The Pareto-efficient solution for minimizing the team cost function (5.1) is achieved by invoking the following strategy:

$$U^*(\alpha) = \arg\min_{U\in\mathscr{U}} \sum_{i=1}^{N} \alpha^i J^i(U) = \arg\min_{U\in\mathscr{U}} J^c(\alpha), \tag{5.3}$$

in which $U^*(\alpha)$ is the optimal value of $U(\alpha)$.

The set of solutions to the minimization problem (5.3) is a function of the parameter α which provides a set of Pareto-efficient solutions. Among these solutions, a unique solution can be obtained by using one of the methods that was mentioned in Chap. 2, namely the *NBS*. Using this method, the unique solution to the problem (a unique α) is given by (2.41) in which J^i's are defined in (3.5) and are calculated by applying the solution of the minimization problem (5.3) to the system that is given in (2.7), and which is therefore a function of the parameter α. The terms d^i's are the values of the cost that is defined in (3.5) which is obtained by applying a noncooperative approach (e.g., a decentralized optimal controller) to the individual subsystems in (2.6).

By solving the maximization problem (2.41), the parameter α can be obtained and substituted into the set of control strategies that are given in (5.3). This solution guarantees that the product of the distances between d^i's (noncooperative solution) and J^i's (cooperative solution) is maximized, implying that the individual costs in the latter case are minimized as much as possible.

Let us first solve the minimization problem (5.3), and then apply a modified version of the algorithm that is given in [26] to solve the maximization problem (2.41).

A preliminary version of the following material has been published in [102, 110, 112].

5.1.2 Solution of the Minimization Problem: An LMI Formulation

In order to solve the minimization problem (5.3), the cost function (5.1) should be minimized subject to the dynamical constraint (2.7). This is a standard *LQR* problem and its solution for an infinite horizon case (i.e., $T \to \infty$) will result in the following control law:

$$U^*(\alpha, X) = -R^{-1}B^T PX, \quad Q - PBR^{-1}B^T P + PA + A^T P = 0. \tag{5.4}$$

The control U^* can be constructed if the above algebraic Riccati equation (ARE) has a solution for P. However, some issues arise when the above control law is applied to the dynamical system (2.7). In fact, given that the matrix P is not guaranteed to be block diagonal, the control signal U^* yields a *centralized strategy* in the sense that its components, i.e., u_l^{i*}, are dependent on the information from the *entire* team. Moreover, the solution suggested by (5.4) does not necessarily guarantee that a nonzero consensus is achieved for an arbitrary parameter selection. In other words, zero consensus is also a possible solution of (5.4).

Hence, to ensure that a desirable consensus solution is obtained that also satisfies the constraints on the availability of the information, one needs to impose additional constraints on the original minimization problem. However, by adding additional constraints to the cost function (5.1), e.g., by considering a barrier function, the problem will no longer be a convex optimization problem and may not necessarily have a unique solution. To remedy this problem, the original cost function is kept unchanged; however, the optimization problem is now formulated as an *LMI* problem so that the constraints due to the consensus and the controller structure are incorporated as convex constraints.

As was pointed out in Chap. 2, the *LQR* problem can be formulated as a maximization or a minimization problem subject to a set of matrix inequalities. In other words, instead of solving the ARE (5.4), e.g., the following maximization problem can be solved:

$$\begin{aligned} &\max \operatorname{trace}(P) \ \text{ s.t.} \\ &PA + A^T P - PBR^{-1}B^T P + Q \geq 0, \ P \geq 0. \end{aligned} \tag{5.5}$$

The above formulation that is provided in (2.36) can be transformed into an *LMI* maximization problem which can be stated as the following problem.

5.1. Problem A

$$\max \operatorname{trace}(P) \ \text{ s.t. } \begin{bmatrix} PA + A^T P + Q & PB \\ B^T P & R \end{bmatrix} \geq 0, \ P \geq 0. \tag{5.6}$$

It can be shown that the above maximization problem has a solution if and only if the following ARE has a solution:

$$Q - PBR^{-1}B^T P^T + PA + A^T P = 0. \tag{5.7}$$

Moreover, if $R > 0$ and $Q \geq 0$, the unique optimal solution to the maximization Problem A is the maximal solution to the ARE in (5.7) [1].

In the above discussion, we have demonstrated how one can formulate the optimization problem as a set of *LMI*s. Solutions to this set of *LMI*s which also minimizes the cost function (5.1) guarantee the consensus seeking, i.e., $X \to \xi\mathbf{1}$. However, among these solutions, a possible solution is when $\xi = 0$, i.e., when the closed-loop system is asymptotically stable and converges to the origin. This solution is not desirable since it is a trivial solution of the consensus-seeking problem and should be avoided. For this purpose, we may add the consensus-seeking condition to Problem A, i.e. $(A - BR^{-1}B^T P)S = 0$ should be incorporated into Problem A. Here, S denotes the unity vector, i.e., $S = \mathbf{1}$. This constraint will guarantee that the closed-loop matrix has a zero eigenvalue and is not Hurwitz. Therefore, if the remaining eigenvalues of this matrix are negative, i.e., it is stable, then the system trajectory will move towards a constant nonzero state which is in the consensus space S. On the other hand, by adding other constraints to the *LMI* problem, as will be discussed subsequently in this section, stability of the closed-loop matrix would be guaranteed as well. We now have a new formulation to our problem which is stated next.

The *LQR* minimization problem for the consensus seeking can be formulated as a maximization problem subject to a set of *LMI*s, namely,

5.2. Problem B

$$\max \operatorname{trace}(P) \text{ s.t.}$$
$$\begin{cases} 1.\ \begin{bmatrix} PA + A^T P + Q & PB \\ B^T P & R \end{bmatrix} \geq 0,\ P \geq 0, \\ 2.\ (A - BR^{-1}B^T P)S = 0, \end{cases} \tag{5.8}$$

where the optimal control law is selected as $U^* = -R^{-1}B^T PX$ and P is obtained by solving the above set of *LMI*s.

5.1.2.1 Consensus-Seeking Subject to a Predefined Information Structure

As discussed previously, the solution to the above problem as well as the one given in (5.4) requires *full* network information for each agent. However, each agent has only access to its neighboring set information. Therefore, one needs to impose a constraint on the controller structure in order to practically satisfy the availability of information. For the sake of notational simplicity assume that each agent has a one-dimensional state-space representation, i.e., A^i in (2.6) is a scalar. The case of a non-scalar system matrix can be treated similarly. The controller coefficient, i.e., $R^{-1}B^T P$ in Problem B should have the same structure as the Laplacian matrix so that the neighboring set constraint holds. However, due to their definitions, both R and B are block diagonal. Therefore, it suffices to restrict P to have the same structure as the Laplacian matrix, i.e., $P(i, j) = 0$ if $L(i, j) = 0$, where $L(i, j)$ designates the ijth entry of the Laplacian matrix L. We may now solve the following problem to minimize the cost function (5.1) while simultaneously satisfying all the problem constraints, namely we now have

5.3. Problem C

$$\max \operatorname{trace}(P) \text{ s.t.}$$

$$\begin{cases} 1. \begin{bmatrix} PA + A^T P + Q & PB \\ B^T P & R \end{bmatrix} \geq 0,\ P \geq 0, \\ 2. (A - BR^{-1}B^T P)S = 0, \\ 3. P(i,j) = 0 \text{ if } L(i,j) = 0,\ \forall i,j = 1,\ldots,N. \end{cases} \tag{5.9}$$

This problem is an *LMI* maximization problem in terms of P. Up to now, we have formulated the minimization problem (5.3) as a set of *LMI*s. Now, let us try to solve the maximization problem that is given by (2.41) [or (2.40)]. For this purpose, we need to calculate the individual "selfish" agent costs J^i by utilizing a given method. For this purpose, the semi-decentralized optimal control strategy that was developed previously in Chap. 3 is used. These values are considered as the *noncooperative* outcome of the team, referred to as d^i's in (2.41) [or (2.40)]. We use the proposed optimal control strategy that results from the individual minimization of the agent cost functions (3.5) as provided in Lemma 3.1 (or Lemma 3.4).

Remark 5.1. It is worth noting that any algorithm which guarantees consensus seeking can be considered as a cooperative algorithm. However, in the context of the formulation based on the game theory, the approach based on the semi-decentralized optimal control is classified as noncooperative. The reason for such designation follows from our previous discussions and definitions where the game theoretic framework yields more characteristics of a cooperative solution when compared to the solution that is obtained by the optimal control strategy. Consequently, the cost values that are obtained using the semi-decentralized optimal control strategy are referred to as the noncooperative outcomes (or threat points).

We are now in a position to develop an algorithm for determining an *NBS* to the cooperative game theory problem.

5.1.3 An Algorithm for Obtaining a Nash Bargaining Solution

Up to this point, we have shown that for any given $\alpha > 0$ the maximization Problem C should first be solved. We now need an algorithm for solving the maximization problem (2.41) [or (2.40)] over different values of α so that a suitable and unique α can be obtained. In [26], two numerical algorithms for solving this maximization problem are given. With minor modifications made to one of these algorithms, the following algorithm will be used for numerical simulation purposes. Namely, we have:

Algorithm I

- **Step 1** Start with an initial selection for $\alpha_0 \in \mathscr{A}$ (e.g., $\alpha_0 = [1/N,\dots,1/N]$ is a good choice).
- **Step 2** Compute $U^*(\alpha_0) = \arg\min_{U\in\mathscr{U}} \sum_{i=1}^{N} \alpha_0^i J^i(U)$ by solving the maximization Problem C.
- **Step 3** Verify if $J^i(U^*) \le d^i$, $i = 1,\dots,N$, where d^i is the optimal value of (3.5) when the control laws (3.15)–(3.17) [or control laws (3.50)–(3.52)] are applied to the dynamical subsystems (2.6). If this condition is not satisfied, then there is at least one i_0 for which $J^{i_0}(U^*) > d^{i_0}$. In that case, update $\alpha_0^{i_0} = \alpha_0^{i_0} + 0.01$, $\alpha_0^i = \alpha_0^i - \frac{0.01}{N-1}$, for $i \neq i_0$ and return to Step 2 (similarly extend the update rule for more than one i_0).
- **Step 4** Calculate

$$\tilde{\alpha}^j = \frac{\prod_{i\neq j}(d^i - J^i(U^*(\alpha_0)))}{\sum_{i=1}^{N}\prod_{k\neq i}(d^k - J^k(U^*(\alpha_0)))},\ j = 1,\dots,N.$$

- **Step 5** Apply the update rule $\alpha_0^i = 0.9\alpha_0^i + 0.1\tilde{\alpha}^i$. If $|\tilde{\alpha}^i - \alpha_0^i| < 0.01$ for $i = 1,\dots,N$, then terminate the algorithm and set $\alpha = \tilde{\alpha}$, else return to Step 2.

The above discussions are now summarized in the following theorem.

Theorem 5.1. *Consider a team of multi-agent systems with individual dynamical representation (2.6) or the team dynamics (2.7), the individual cost function (3.5), and the team cost function (5.1) with the corresponding parameters (5.2). Furthermore, assume that the desirable value of the parameter α is found by using the Algorithm I. Moreover, the control law U^* is designed as $U^* = -R^{-1}B^T PX$, and P is the solution to the following optimization problem:*

$$\max \operatorname{trace}(P) \text{ s.t.}$$

$$\begin{cases} 1.\ \begin{bmatrix} PA + A^T P + Q & PB \\ B^T P & R \end{bmatrix} \ge 0,\ P \ge 0, \\ 2.\ A_c = (A - BR^{-1}B^T P),\ A_c S = 0, \\ 3.\ P(i,j) = 0 \text{ if } L(i,j) = 0,\ \forall i,j = 1,\dots,N, \end{cases} \tag{5.10}$$

where $S = \mathbf{1}$. It then follows that

(a) In the infinite horizon scenario, i.e., $T \to \infty$, the above controller solves the following min-max problem:

$$U^* = \arg\min_{U\in\mathscr{U}} \sum_{i=1}^{N} \alpha^i J^i(U),\ \alpha \in \mathscr{A},$$

$$\mathscr{A} = \left\{ \alpha = (\alpha^1,\dots,\alpha^N) | \alpha^i \ge 0 \text{ and } \sum_{i=1}^{N} \alpha^i = 1 \right\},$$

$$\alpha^* = \arg\max_{\alpha} \prod_{i=1}^{N} (d^i - J^i(\alpha, U^*)),\ J \preceq d.$$

The solution to the above min–max problem guarantees consensus achieving for the proposed team of multi-agent systems, that is, in the steady state $X \to \xi \mathbf{1}$, where ξ is a constant coefficient of the consensus value.

(b) *In addition, the suggested control law guarantees a stable consensus of agents output to a common value subject to the dynamical and information structure constraints of the team in a cooperative manner, if for at least one connected subgraph of the original graph, we have*

$$A_c(i,j) \neq 0 \text{ if } L_{\text{sub}}(i,j) \neq 0, \ \forall i,j = 1,\ldots,N, \tag{5.11}$$

where L_{sub} denotes the Laplacian matrix of any such arbitrary connected subgraph.

(c) *Moreover, the optimal value of the cost function (5.1) has a finite infimum of $X^T(0)PX(0) - \xi^2 \sum_i \sum_j P(i,j)$, where P is obtained from (5.10) and $X(0)$ is the initial value of the entire team state vector.*

Proof. The proof is provided in Appendix A. □

We can now conclude that by using the above results, the team consensus goal can be achieved in a decentralized cooperative manner while simultaneously satisfying all the given team information constraints.

5.2 An LMI Approach to the Optimal Consensus Seeking

In this section, an optimal consensus protocol is designed by using the optimal control and the *LMI* design tools. For this purpose, the idea of decomposing the state vector into two components, as introduced in [24], is adopted for solving the optimal consensus problem. As opposed to [24], where H_2 design methodology is used for design of a robust consensus-seeking algorithm, here we start with a Hamilton–Jacobi–Bellman equation. We will then show the difficulties that arise when this formulation is utilized. Therefore, we propose the LMI formulation of the LQR problem. After decomposing the state vector, a global cost function is suggested for the entire network to achieve a stable consensus. Minimization of this global cost function guarantees a stable consensus with an optimal control effort. The global cost function formulation provides more insight into the optimal performance of the entire network and would result in a global optimal (or suboptimal) solution.

In what follows, we first decompose the state vector of the entire team into components in the consensus subspace and its orthogonal subspace. This decomposition helps to reduce the consensus-seeking problem into a stabilization problem. This stabilization problem is then formulated and solved by using the optimal control technique and the *LMI*s.

Remark 5.2. For only this section, A^* denotes the complex conjugate transpose of A, whereas in the rest of this book, A^* denotes the optimal value of the quantity A.

A preliminary version of the material presented below has been published in [111, 114].

5.2.1 State Decomposition

Using the consensus definition that is given in Definition 2.6, the orthonormal basis for the subspace $\mathscr{S}$ is denoted by $S_{Nn\times 1} = \mathbf{1}$. The orthonormal complement of this matrix is denoted by $\bar{S}_{Nn\times(Nn-1)}$, which is a basis for the corresponding subspace orthonormal to $\mathscr{S}$. The following relationships are satisfied by these matrices:

$$\bar{S}^*S = 0,\ \bar{S}^*\bar{S} = I,\ S^*S = 1,\ \bar{S}\bar{S}^* + SS^* = I. \tag{5.12}$$

Now, the state vector X can be decomposed into two orthogonal components in the above-mentioned subspaces and can be written as [24]

$$X = [\bar{S}\ S]\begin{bmatrix} X_{\bar{s}} \\ X_s \end{bmatrix}. \tag{5.13}$$

Assuming that the control input has a state feedback structure, i.e., $U = KX$, then the dynamical equation of the system given by (2.7) will be transformed into

$$\begin{bmatrix} \dot{X}_{\bar{s}} \\ \dot{X}_s \end{bmatrix} = \begin{bmatrix} \bar{S}^* \\ S^* \end{bmatrix}(A+BK)[\bar{S}\ S]\begin{bmatrix} X_{\bar{s}} \\ X_s \end{bmatrix}. \tag{5.14}$$

This follows from the fact that $[\bar{S}\ S]^{-1} = \begin{bmatrix} \bar{S}^* \\ S^* \end{bmatrix}$.

Since the goal is to ensure consensus in the subspace $\mathscr{S}$ for the closed-loop system, the following equilibrium condition is imposed on the above dynamical equation:

$$(A+BK)S = 0. \tag{5.15}$$

In other words, the equilibria should lie in the consensus subspace. This condition should then be incorporated in the design procedure. Therefore, we will have

$$\begin{bmatrix} \dot{X}_{\bar{s}} \\ \dot{X}_s \end{bmatrix} = \begin{bmatrix} \bar{S}^* \\ S^* \end{bmatrix}(A+BK)\bar{S}X_{\bar{s}} = \begin{bmatrix} \bar{S}^*(A+BK)\bar{S}\ 0 \\ S^*(A+BK)\bar{S}\ 0 \end{bmatrix}\begin{bmatrix} X_{\bar{s}} \\ X_s \end{bmatrix}. \tag{5.16}$$

In order to achieve consensus, the final state of the system should be a vector in the subspace $\mathscr{S}$. Therefore, the component $X_{\bar{s}}$ should converge to zero in steady state. This implies that this part of the system dynamics should be asymptotically stable. Moreover, the dynamics corresponding to X_s is only dependent on $X_{\bar{s}}$, and therefore we are only concerned with the dynamics corresponding to $X_{\bar{s}}$ as governed by

$$\dot{X}_{\bar{s}} = \bar{S}^*(A+BK)\bar{S}X_{\bar{s}} = \bar{S}^*A\bar{S}X_{\bar{s}} + \bar{S}^*BK\bar{S}X_{\bar{s}} = \bar{A}X_{\bar{s}} + \bar{B}\bar{K}X_{\bar{s}} = \bar{A}X_{\bar{s}} + \bar{B}\bar{U}, \tag{5.17}$$

where

$$\bar{A} = \bar{S}^*A\bar{S},\ \bar{B} = \bar{S}^*B,\ \bar{K} = K\bar{S},\ \bar{U} = \bar{K}X_{\bar{s}}. \tag{5.18}$$

If this part of the dynamics is stabilized asymptotically to zero, $X_{\bar{s}}$ will approach to zero, and hence X_s will lead to a constant value. On the other hand, if condition (5.15) is imposed, this constant value will be in the consensus subspace. Therefore, the consensus will be achieved.

We may now design a state feedback control strategy to guarantee the consensus achievement corresponding to the closed-loop system. Towards this end, optimal control techniques will be used below to design the controller to guarantee a stable consensus in an optimal manner. As already indicated in the above discussion, the purpose of the control design is to stabilize the portion of the system dynamics that corresponds to the subspace $\bar{\mathscr{S}} = \text{span}\{\bar{S}\}$. Therefore, our objective is to design the control gain, namely $\bar{K}$. Based on this value of $\bar{K}$, the associated value of K for the original system can be obtained. In the following, this controller is designed by using optimal control techniques.

5.2.2 *Optimal Control Design*

Optimality here refers to the situation when the dynamics of $X_{\bar{s}}$ is stabilized in an optimal manner. For characterizing optimality we need to define a formal performance index. We can define either individual performance indices or a single index (cost function) for the entire team. In Chap. 3, we proposed individual cost functions and suggested a semi-decentralized control strategy for minimizing these cost functions. Although, the individual cost functions can achieve better fit within a decentralized control structure, they cannot be utilized as an index for the team performance. In contrast, the team cost function which is used here is a good index for the team performance and its minimization can result in a globally optimal (or suboptimal) solution. However, the solution will be centralized. Fortunately, by using the *LMI* formulation, it will be shown that this centralized solution can be avoided by adding a constraint on the structure of the controller gain matrix.

In order to stabilize the dynamics that is given in (5.17), let us define the team cost function that is to be minimized as follows:

$$d = \int_0^\infty \{X_{\bar{s}}^T \hat{Q} X_{\bar{s}} + \bar{U}^T R \bar{U}\} \mathrm{d}t, \; X = [\bar{S} \; S] \begin{bmatrix} X_{\bar{s}} \\ X_s \end{bmatrix}, \tag{5.19}$$

where $\hat{Q}$ has a predefined structure as $\hat{Q} = \bar{S}^* Q \bar{S} > 0$ and Q and R are the *PD* matrices (if $Q_{Nn \times Nn}$ is selected to be a *PD* matrix, since $\text{rank}(\bar{S}_{(Nn)\times(Nn-1)}) = Nn - 1$, then $\hat{Q}$ will also be a PD matrix [59]).

In the following, we will show that in general the minimization of this cost function using the Riccati equation does not result in a consensus for a network of multi-agent systems with general dynamical representation. Therefore, in the following sections, an *LMI* formulation is utilized for the optimization problem which can incorporate the requirements of the consensus achievement and results in an optimal consensus algorithm.

5.2.2.1 Discussion on the Solution of the Riccati Equation

The problem of minimizing the cost function (5.19) subject to the dynamical constraint (5.17) is a standard *LQR* problem. The solution to this *LQR* problem can be achieved by solving the corresponding Riccati equation as follows:

$$\bar{U} = -R^{-1}\bar{B}^*PX_{\bar{S}}, \tag{5.20}$$

where P satisfies the following Riccati equation:

$$P\bar{A} + \bar{A}^*P - P\bar{B}R^{-1}\bar{B}^*P + \hat{Q} = 0. \tag{5.21}$$

Therefore, $\bar{K} = K\bar{S} = -R^{-1}\bar{B}^*P$ and from the properties of the matrix $\bar{S}$ one can obtain K as $K = -R^{-1}\bar{B}^*P\bar{S}^*$. Hence, the control input to the original system is given by

$$U = -R^{-1}\bar{B}^*P\bar{S}^*X. \tag{5.22}$$

By applying this input to the system (2.7), the closed-loop dynamics can be written as

$$\dot{X} = (A - BR^{-1}\bar{B}^*P\bar{S}^*)X. \tag{5.23}$$

In order to achieve consensus for the closed-loop system, the matrix S should be in the null-space of the closed-loop matrix, i.e., $[A - BR^{-1}\bar{B}^*P\bar{S}^*]S = 0$. However, the second part of this expression is zero due to the properties of $\bar{S}$ as stated in (5.12), i.e.,

$$-BR^{-1}\bar{B}^*P\bar{S}^*S = 0. \tag{5.24}$$

In other words, we should have $AS = 0$ to guarantee a stable consensus. The above discussion is formally summarized in the following lemma.

Lemma 5.1. *Consider a team of multi-agent systems with its entire dynamics given by (2.7), where interaction terms are incorporated in the agents' dynamics. Assume that a state decomposition procedure is performed and the consensus-seeking problem is reduced to the stabilization of the dynamical equation (5.17). Then, the solution of the corresponding Riccati equation given by (5.21) which minimizes the cost function (5.19) subject to the dynamical constraint (5.17) may not result in a stable consensus algorithm unless the consensus subspace is in the null space of the open-loop matrix A, i.e.,*

$$AS = 0. \tag{5.25}$$

In other words, this solution may not provide a stable equilibria in the consensus subspace.

Proof. Follows from the previous constructive results. □

In general, the condition $AS = 0$ may not be satisfied by the subsystems in the network. Therefore, for a system with an arbitrary matrix A, the optimal solution obtained by solving the Riccati equation does not necessarily guarantee consensus achievement. To simply explain this observation, one may note that according to the definition of the control U given in (5.22), this signal only provides a component in $\bar{\mathscr{S}}$. Hence, the term BU does not contribute to the $\mathscr{S}$ component of $\dot{X}$. Therefore, to have a stable solution where $\dot{X} = 0$, the term AX should enjoy the same property, i.e., the component of AX in $\mathscr{S}$ subspace should be zero:

$$AS = 0. \tag{5.26}$$

Since this condition is not generally satisfied for an arbitrary system matrix A, the consensus condition (5.15) in general should be imposed onto the optimal solution that is achieved through the solution of the proposed minimization problem as an extra constraint. Hence, instead of obtaining an optimal solution through the solution of the Riccati equation, in the following, we try to obtain an optimal solution for the above minimization problem subject to the consensus constraint through the solution of a set of *LMI*s.

5.2.2.2 LMI Formulation of the Optimal Consensus Seeking

As in the previous section, the problem of minimizing the cost function (5.19) subject to the dynamical constraint (5.17) cannot be solved as a standard *LQR* problem if the consensus seeking is to be added as a constraint. Instead, we may use one of the *LMI* formulations for solving the optimal problem which was introduced in Chap. 2. In other words, instead of solving the ARE (5.21), the controller $\bar{U} = \bar{K}X_{\bar{S}}$ that minimizes the cost function (5.19) subject to (5.17) is achieved by solving for and determining the appropriate matrix P according to

$$\begin{aligned} &\min \operatorname{trace}(P) \ \text{ s.t.} \\ &P(\bar{A}+\bar{B}\bar{K}) + (\bar{A}+\bar{B}\bar{K})^*P + \hat{Q} + \bar{K}^*R\bar{K} \leq 0, \ P \geq 0, \end{aligned} \tag{5.27}$$

where $\bar{K} = -R^{-1}\bar{B}^*P$ yields the optimal solution. For the dynamical system (5.17) the inequality constraint (5.27) can be written as

$$P(\bar{S}^*A\bar{S} + \bar{S}^*BK\bar{S}) + (\bar{S}^*A\bar{S} + \bar{S}^*BK\bar{S})^*P + \bar{S}^*Q\bar{S} + \bar{S}^*K^*RK\bar{S} \leq 0. \tag{5.28}$$

By multiplying both sides of this inequality by P^{-1} we get

$$\bar{S}^*(A+BK)\bar{S}P^{-1} + P^{-1}\bar{S}^*(A^*+K^*B^*)\bar{S} + P^{-1}\bar{S}^*Q\bar{S}P^{-1} + P^{-1}\bar{S}^*K^*RK\bar{S}P^{-1} \leq 0. \tag{5.29}$$

Now define a new variable $Z = Z^* > 0$ that satisfies the following equation [24]:

$$Z = \bar{S}\bar{S}^*Z\bar{S}\bar{S}^* + SS^*ZSS^*, \tag{5.30}$$

an example of which can be in the following form [24]:

$$Z = [\bar{S}\ S] \begin{bmatrix} P^{-1} & 0 \\ 0 & M \end{bmatrix} \begin{bmatrix} \bar{S}^* \\ S^* \end{bmatrix}, \tag{5.31}$$

where $M = S^*ZS$ and $P^{-1} = \bar{S}^*Z\bar{S}$ are PD matrices and P can be the same matrix as the one used in (5.29). Corresponding to this definition of Z we will have $Z\bar{S} = \bar{S}P^{-1}$. Substituting $Z\bar{S} = \bar{S}P^{-1}$ into (5.29) one gets

$$\bar{S}^*(AZ + BKZ + ZA^* + ZK^*B^* + ZQZ + ZK^*RKZ)\bar{S} \leq 0. \tag{5.32}$$

Let us now introduce a new variable $W = KZ$, such that one gets

$$\bar{S}^*(AZ + BW + ZA^* + W^*B^* + ZQZ + W^*RW)\bar{S} \leq 0. \tag{5.33}$$

This can be written as an *LMI* condition using the Schur complement and noting that $R > 0$ and $Q \geq 0$, namely

$$\begin{bmatrix} \Upsilon & \bar{S}^*ZQ^{1/2} & \bar{S}^*W^*R^{1/2} \\ Q^{1/2}Z\bar{S} & -I & 0 \\ R^{1/2}W\bar{S} & 0 & -I \end{bmatrix} \leq 0,\ \Upsilon = \bar{S}^*(AZ + BW + ZA^* + W^*B^*)\bar{S}, \tag{5.34}$$

where Z, W are *LMI* parameters. Therefore, we have shown that the minimization problem in (5.27) can be written as follows:

$$\min \operatorname{trace}(P) \ \text{s.t.}$$

$$\begin{bmatrix} \Upsilon & \bar{S}^*ZQ^{1/2} & \bar{S}^*W^*R^{1/2} \\ Q^{1/2}Z\bar{S} & -I & 0 \\ R^{1/2}W\bar{S} & 0 & -I \end{bmatrix} \leq 0,$$

$$\Upsilon = \bar{S}^*(AZ + BW + ZA^* + W^*B^*)\bar{S},\ \ Z = \bar{S}\bar{S}^*Z\bar{S}\bar{S}^* + SS^*ZSS^*,\ \ Z\bar{S} = \bar{S}P^{-1}, \tag{5.35}$$

where $K = WZ$. In the following, we will discuss the conditions for existence of a solution to the above minimization problem and then present the main results of this section as a theorem.

5.2.2.3 Discussion on the Existence of Solutions

It is well-known that detectability and stabilizability conditions are sufficient for existence of a unique stabilizing solution to a linear quadratic optimal control problem. The following lemma illustrate and formulate these conditions for our specific problem.

Lemma 5.2. *The minimization problem (5.27), or equivalently (5.35), subject to the dynamical constraint (5.17) has an optimal stabilizing solution if the matrices A, B, and Q are given such that the following inequalities have a solution for P_2:*

1. *Stabilizability condition:*

$$\bar{S}^*(AP_2 + P_2A^* - BB^*)\bar{S} < 0, \tag{5.36}$$

2. *Detectability condition:*

$$\bar{S}^*(P_2A + A^*P_2 - Q)\bar{S} < 0, \tag{5.37}$$

where $P_2 > 0$ satisfies $P_2 = \bar{S}\bar{S}^*P_2\bar{S}\bar{S}^* + SS^*P_2SS^*$.

Proof. The details are given in Appendix A. □

Remark 5.3. For our present problem, the system matrix A given by (2.8) is a function of the interaction coefficients, $\mathscr{F}^{ij}$, and therefore can be viewed as a design parameter. In case that the matrices B, Q, and the matrix A with no interaction terms, i.e., $\mathscr{F}^{ij} = 0$, $\forall i, j$, satisfy the conditions (5.36) and (5.37), the existence of a solution is guaranteed. However, if these conditions are not guaranteed, then we may select the coefficients $\mathscr{F}^{ij}$ to satisfy them. A simple approach is to take P_2 as an identity matrix and then select A such that both inequalities are satisfied. However, it should be noted that in some special conditions and due to the special structure of the matrix A, it might not be possible to obtain a solution for the inequalities in (5.36), (5.37) by using this method. An example of this situation is when $A^i = 0$, $\forall i$, and Q has the same structure as the Laplacian matrix. In this case $A + A^* - Q$ will have a positive eigenvalue regardless of the selection of the parameters of the matrix A. In this situation one possible solution would be to add an internal loop for each individual controller so that the required conditions in (5.36) and (5.37) are satisfied (by adding diagonal elements to the matrix A).

The main result of this section and the conclusion from the above discussions are now summarized in the following theorem.

Theorem 5.2. *(a) Consider a team of multi-agent systems with the dynamical equation as in (2.7). Assume that the state vector is decomposed into components in the consensus subspace and its orthogonal subspace. Therefore, the dynamics corresponding to this decomposition is given by (5.17) with the cost function given by (5.19) is to be minimized for stabilization of this dynamics. Also, assume that the matrices A, B, and Q satisfy the conditions of Lemma 5.2. Moreover, assume that the control input U is selected as $U = KX$, where $K = WZ^{-1}$ and the* LMI *variables Γ, W, and Z are obtained through the minimization problem below:*

$$\min \operatorname{trace}(\Gamma) \text{ s.t.}$$

$$\begin{cases} 1. \begin{bmatrix} \Gamma & I \\ I & \bar{S}^* Z \bar{S} \end{bmatrix} > 0, \\ 2. \begin{bmatrix} \Upsilon & \bar{S}^* Z Q^{1/2} & \bar{S}^* W^* R^{1/2} \\ Q^{1/2} Z \bar{S} & -I & 0 \\ R^{1/2} W \bar{S} & 0 & -I \end{bmatrix} \leq 0, \Upsilon = \bar{S}^*(AZ + BW + ZA^* + W^* B^*)\bar{S}, \\ 3. (AZ + BW)S = 0, \\ 4. Z = \bar{S}\bar{S}^* Z \bar{S}\bar{S}^* + SS^* Z SS^*, \ Z > 0 \end{cases} \tag{5.38}$$

Then, the cost function (5.19) is minimized and the system (5.17) is asymptotically stabilized. This in turn makes the system in (2.7) reach a stable consensus in an optimal manner.

(b) Furthermore, if the following constraints are added to the above minimization problem, the controller will be semi-decentralized. In other words, only partial information that is available through the predefined neighboring sets will be used by individual controllers, provided that

$$\begin{cases} 1. \, Z \text{ is diagonal, i.e. } Z = \begin{pmatrix} Z_1 & \dots & 0 \\ \vdots & \ddots & \vdots \\ 0 & \dots & Z_N \end{pmatrix}, \text{ and} \\ 2. \, W(i,j) = 0 \text{ if } L(i,j) = 0, \end{cases} \tag{5.39}$$

where L is the Laplacian matrix of the graph describing the network.

Proof. The details are given in Appendix A. □

5.2.3 *Discussion on the Graph Connectivity*

In contrast to most of the consensus-seeking approaches, in the proposed approach in this section, there has been no explicit restriction on the connectivity of the network underlying graph. In the following, we will show that the graph connectivity is a requirement to guarantee consensus achievement. First, we prove the following lemma which is required for the remainder of the discussion in this section.

Lemma 5.3. *The closed-loop matrix of the entire network, i.e., $A + BK$ represents the Laplacian matrix of a weighted graph. The corresponding graph is a subgraph of the original network graph but with different weights assigned to its edges.*

Proof. The details are given in Appendix A. □

Remark 5.4. From the above lemma, it follows that even if the original graph is connected, one may not conclude that the matrix $A + BK$ represents a connected graph. However, in the following, it is shown that for guaranteeing the existence of a solution to the consensus-seeking problem, not only the original graph should be connected but also $A+BK$ should represent the Laplacian of a connected graph.

Theorem 5.3. *(a) If the graph corresponding to the entire network is not connected, the existence of a solution to the consensus problem cannot be guaranteed.*

(b) Moreover, if the consensus seeking is guaranteed, the matrix $A+BK$ will describe the Laplacian of a connected subgraph of the original graph.

Proof. The details are given in Appendix A. □

5.3 Simulation Results

5.3.1 Game Theory Approach

The simulation results that are presented in this section correspond to a team of four agents that are being controlled by using two control strategies, namely the semi-decentralized optimal control law that is given by Lemma 3.1 (or Lemma 3.4) and the cooperative game theoretic-based control law that is given by Theorem 5.1. The first set of numerical simulations corresponds to the application of the control laws (3.15)–(3.17) [or (3.50)–(3.52)] to the individual agent model (2.6). In the second set, the numerical simulation results are obtained by applying the control law $U = KX$ with $K = -R^{-1}B^T P$ to the team dynamics that is described by (2.7). It is assumed that the state vector X^i of each agent is the velocity vector of that agent, i.e., $X^i = v^i$ and that the velocity vector has two components, i.e., $v^i = [v^i_x,\ v^i_y]^T$. The matrix K is obtained by solving the Problem C and by utilizing the maximization Algorithm I. The results that are shown below are conducted through Monte Carlo simulation runs to capture the average behavior of the proposed control strategies. The average team responses are due to 15 different randomly selected initial conditions.

The simulation parameters for both control approaches are selected as follows: $A^i = 0_{2\times 2}$, $R^i = I_{2\times 2}$, $C^i = I_{2\times 2}$, $B^i = \begin{pmatrix} 4 & -3 \\ -2 & 3 \end{pmatrix}$, and $Q^{ij} = \begin{pmatrix} 10 & 3 \\ 3 & 4 \end{pmatrix}$. The random initial conditions of the velocity vector, i.e., v^i_0, for the Monte Carlo simulations are considered as $v^1_0 = [r(6,8)\ \ r(1,3)]^T$, $v^2_0 = [r(5,7)\ \ r(3,5)]^T$, $v^3_0 = [r(2,4)\ \ r(1,3)]^T$, and $v^4_0 = [r(-5,-3)\ \ r(-4,-2)]^T$, where $r(x,\ y)$ represents a random variable in the interval $[x,y]$.

In the cooperative game theory strategy the initial value of the parameter α is selected as $\alpha_0 = [1/4,\ldots,1/4]$, and its optimal average value is obtained by using the procedure that is outlined in Algorithm I for the 15 Monte Carlo simulation runs as $\alpha = [0.2276\ 0.2005\ 0.2486\ 0.3232]$. The interaction gains are selected as $\mathcal{F}^{ij} = 1.6 I_{2\times 2}$.

Table 5.1 A comparative evaluation of the average values of the performance index corresponding to the two control design strategies for the cost functions defined in (3.5) for $T = 2$ s

	Average performance index			
Control scheme	Agent 1	Agent 2	Agent 3	Agent 4
Semi-decentralized optimal control	14,648	14,781	11,629	12,519
Cooperative game theory	8,545	7,730	6,138	8,370

Table 5.1 compares the average values of the cost function (3.5) that are obtained by running the Monte Carlo simulations for the four agents under the two proposed control approaches for a period of 2 s. As expected the average costs for the cooperative game theory approach are less than those that are obtained from the optimal control approach. However, it should be noted that this achievement is at the expense of an increased computational complexity. In fact, in the former method two optimization problems, namely a maximization and a minimization problem should be solved as compared to the semi-decentralized approach where only a single minimization problem needs to be solved. Therefore, there is a trade-off between the control computational complexity and the achievable control performance. A quantitative evaluation criteria of the trade-offs to a large extent will depend on the specific application under investigation and the practical constraints of the system.

In Figs. 5.1a and 5.2a the x-components and in Figs. 5.1b and 5.2b, the y-components of the average velocity profiles of the four-agent team are shown for the semi-decentralized optimal strategy and the game theoretic-based strategy, respectively. Similarly, in Fig. 5.3a, b, the x-component of the average control input efforts of the four-agent team are shown for the semi-decentralized optimal and the game theoretic-based controllers, respectively.

Remark 5.5. It should be noted that the final values that are obtained for the semi-decentralized optimal control strategy are the average of the states' initial values. In fact, the control law provided in Theorem 3.1 is a weighted average consensus protocol which results in consensus on the average value of the initial state vector. However, the cooperative game theory protocol that is obtained by solving the set of *LMI*s given in (5.10) is not necessarily an average consensus protocol. Therefore, in this case, the consensus can be any arbitrary value.

5.3.2 LMI-Based Optimal Control Approach

Remark 5.6. It is worth noting that in the LMI-based optimal control approach, there is more flexibility in design of both the local controllers and the interaction terms. In other words, u_l^i in (2.6) can be a function of both the local information X^i as well as the global information X^j, $j \in N^i$. Consequently, the interaction term $\mathcal{F}^{ij}$ can be selected as zero even though the agent j is in the neighboring set of the agent i.

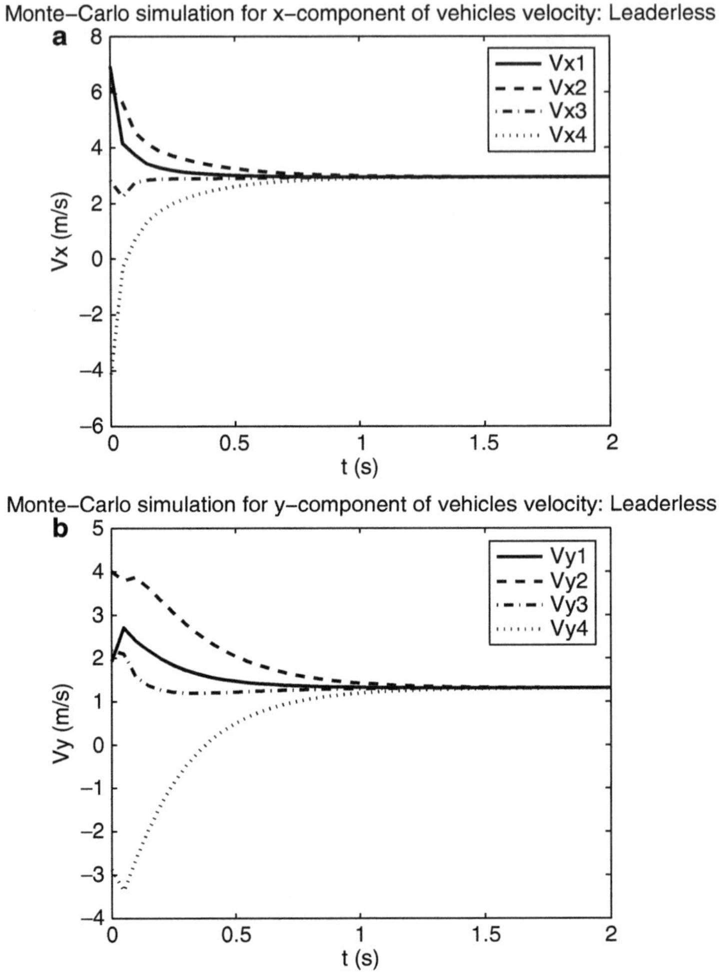

Fig. 5.1 **(a)** The x-component and **(b)** the y-component of the average velocity profiles that are obtained by applying the semi-decentralized optimal control strategy to a team of four agents

In the remainder of this section examples corresponding to two team configurations are presented and simulations are provided.

Example 1. Simulation results conducted for this example are for a team of four agents with the team dynamical model that is given by (2.7) and (2.8). The simulations correspond to two cases. In the first case the requirement that is given in Lemma 5.1 is not satisfied by the system matrix A, i.e., $AS \neq 0$. The Laplacian matrix corresponding to the connected graph describing the network

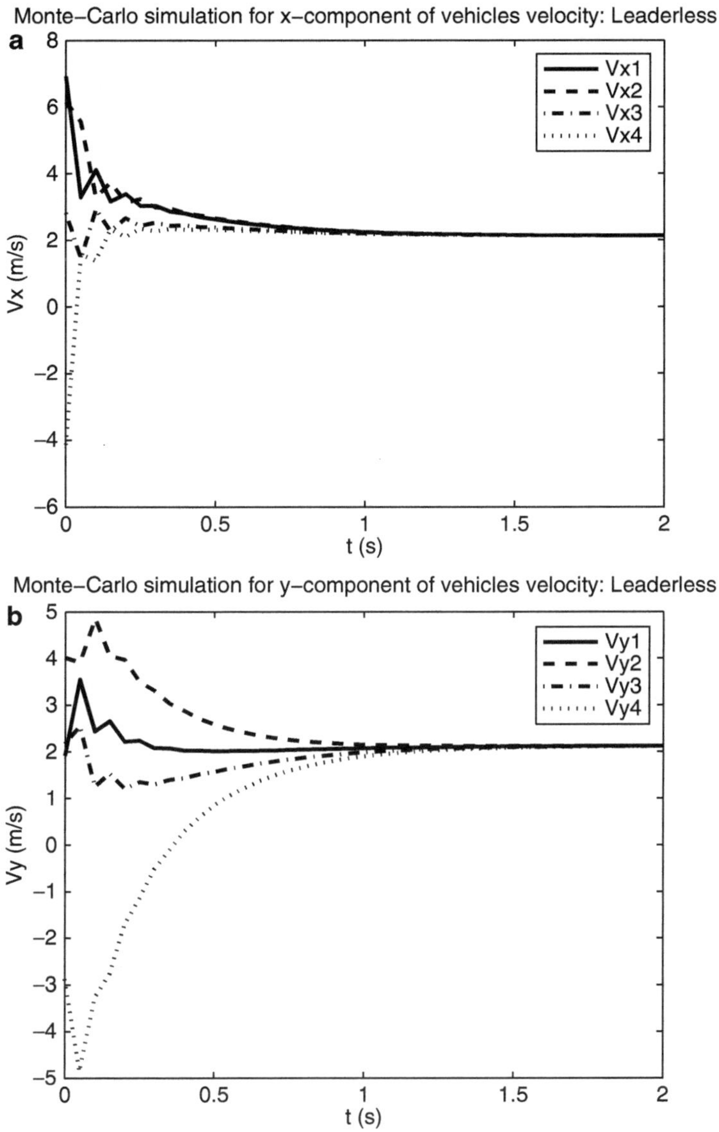

Fig. 5.2 **(a)** The x-component and **(b)** the y-component of the average velocity profiles that are obtained by applying the cooperative game theory strategy to a team of four agents

structure is $L = \begin{pmatrix} 2 & -1 & 0 & -1 \\ -1 & 2 & -1 & 0 \\ 0 & -1 & 2 & -1 \\ -1 & 0 & -1 & 2 \end{pmatrix}$. The other simulation parameters are selected

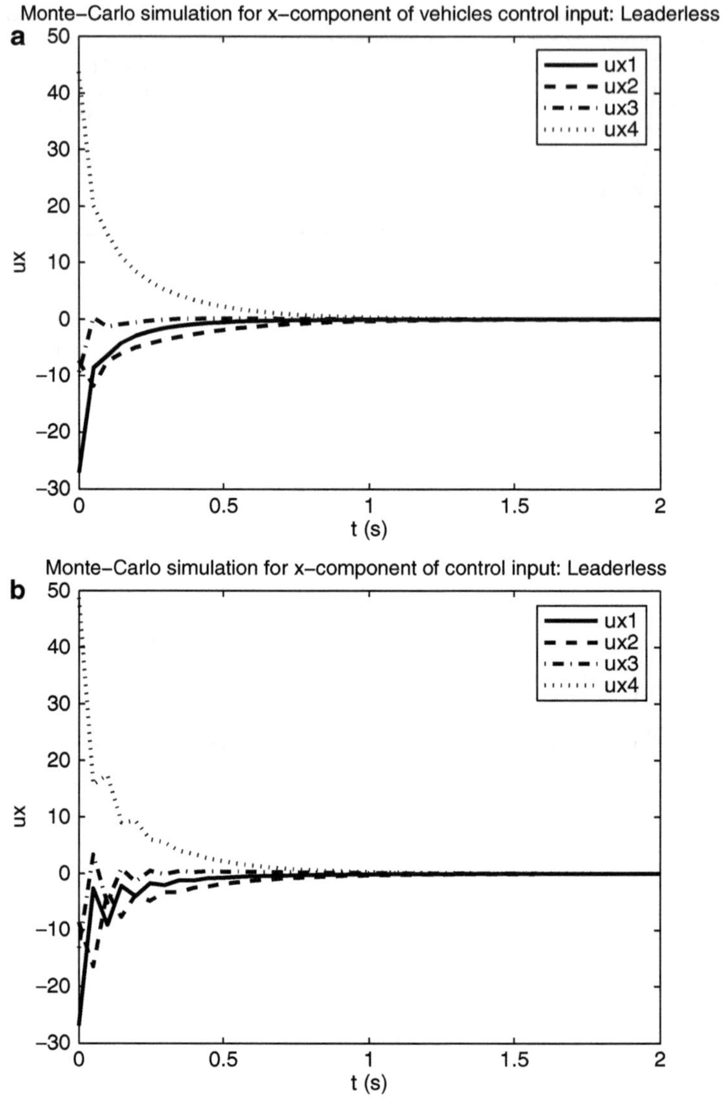

Fig. 5.3 The x-component of the average control efforts that are obtained by applying **(a)** the semi-decentralized optimal control strategy and **(b)** the cooperative game theory approach to a team of four agents

as $A^i = \begin{pmatrix} 1 & 1 \\ 0.5 & 1 \end{pmatrix}$, $B^i = 2I_{2\times 2}$, $c^i = I_{2\times 2}$, $Q^{ij} = 6I_{2\times 2}$, and $R^i = 2I_{2\times 2}$. The initial condition of the state vector is selected as $X(0) = \begin{pmatrix} 6 & 1 & 5 & 3 & 2 & 1 & -5 & -4 \end{pmatrix}^T$. The state vector X is composed of the state vectors of all the agents in the team and the state vector X^i of each agent is the velocity vector of that agent, i.e., $X^i = v^i$ which has two

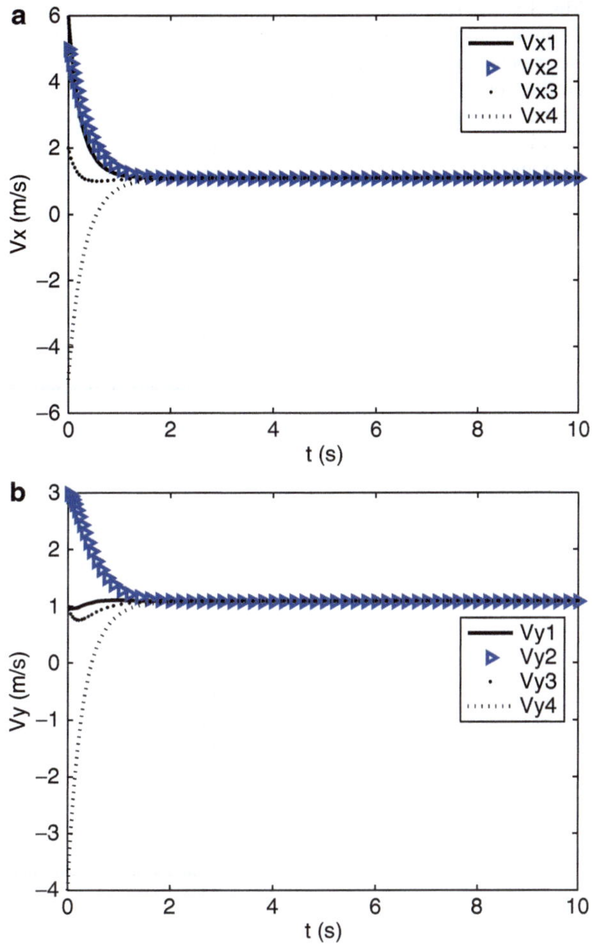

Fig. 5.4 **(a)** The x-component and **(b)** the y-component of the velocity profiles; optimal design based on the solution of the *LMI*s when $AS \neq 0$ in Example 1

components, i.e., $v^i = [v_x^i,\ v_y^i]^T$. The interaction coefficients $\mathscr{F}^{ij}$, $\forall i, j$, are assumed to be zero.

It can be verified that the above parameters satisfy the conditions that are provided in Lemma 5.2, and the corresponding matrix P_2 can be selected as the identity matrix. The simulation results are obtained by applying the control law $U = KX$ to the system (2.7). The matrix K is first evaluated through the set of *LMI*s given in Theorem 5.2. In Fig. 5.4a, the x-component and in Fig. 5.4b, the y-component of the velocity profiles of the four-agent team are shown for the above configuration. In Fig. 5.5a, b we have applied the control strategy that is obtained through the solution of the Riccati equation and given by (5.22) to the system (2.7) with the above configuration. It can be seen that as predicted in Lemma 5.1 the closed-loop system is unstable. This is due to the fact that the open-loop matrix A does not satisfy the property $AS = 0$. Therefore, the closed-loop dynamics cannot

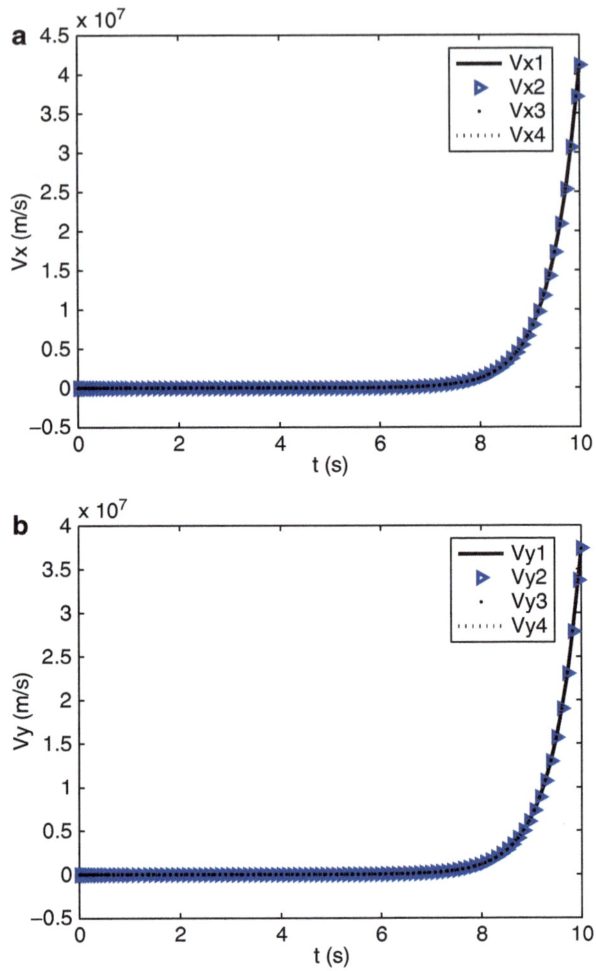

Fig. 5.5 **(a)** The x-component and **(b)** the y-component of the velocity profiles; optimal design based on the solution of the Riccati equation when $AS \neq 0$ in Example 1

provide a stable consensus by using the matrix K that is obtained from the standard LQR-based design methodology by using the Riccati equation.

In the second case of simulations, we have selected a matrix A such that $AS = 0$. The corresponding results are presented in Figs. 5.6 and 5.7. To guarantee the condition $AS = 0$, the interaction coefficients are selected as $\mathscr{F}^{ij} = -0.25 \begin{pmatrix} 1 & 1 \\ 0.5 & 1 \end{pmatrix}$, $\forall i, j \in N^i$ for this case. Other simulation parameters are the same as in the previous case. The simulation results are obtained by applying the controllers that are designed based on the Riccati equation solutions and the solutions to the set of *LMI*s in (5.38). Figure 5.6a, b correspond to the latter approach, whereas Fig. 5.7a, b correspond to the former approach. For comparison between these two approaches we have evaluated a performance index corresponding to both methods.

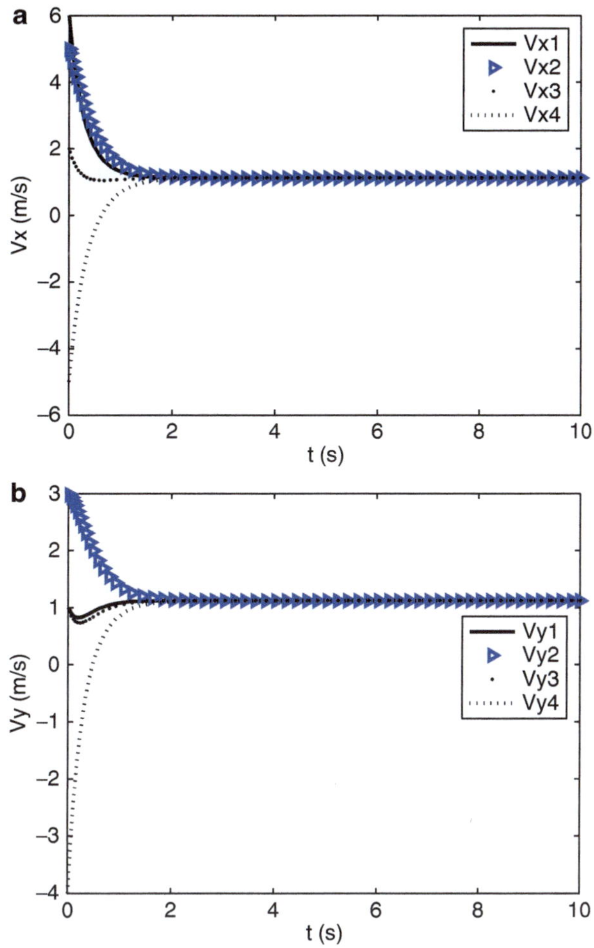

Fig. 5.6 **(a)** The x-component and **(b)** the y-component of the velocity profiles; optimal design based on the solution of the *LMI*s when $AS = 0$ in Example 1

Since one does not have direct access to $X_{\bar{s}}$, instead of the performance index (5.19), we have used the following cost function for comparison purposes, namely

$$\mathrm{PI} = \int_0^T \{X^T QX + U^T RU\} \mathrm{d}t, \tag{5.40}$$

where T is selected as 10 s. The values that are obtained for the above performance index are 865.5 and 883.2 corresponding to the Riccati equation and the *LMI*s approaches, respectively. Also, the controller provided by the solution of the Riccati equation reaches the consensus faster when compared to the controller corresponding to the *LMI*s solution (1.8s vs. 2.6s).

Example 2. The simulation results conducted for this example are presented for a team of six vehicles. The simulations are performed for the two case scenarios that

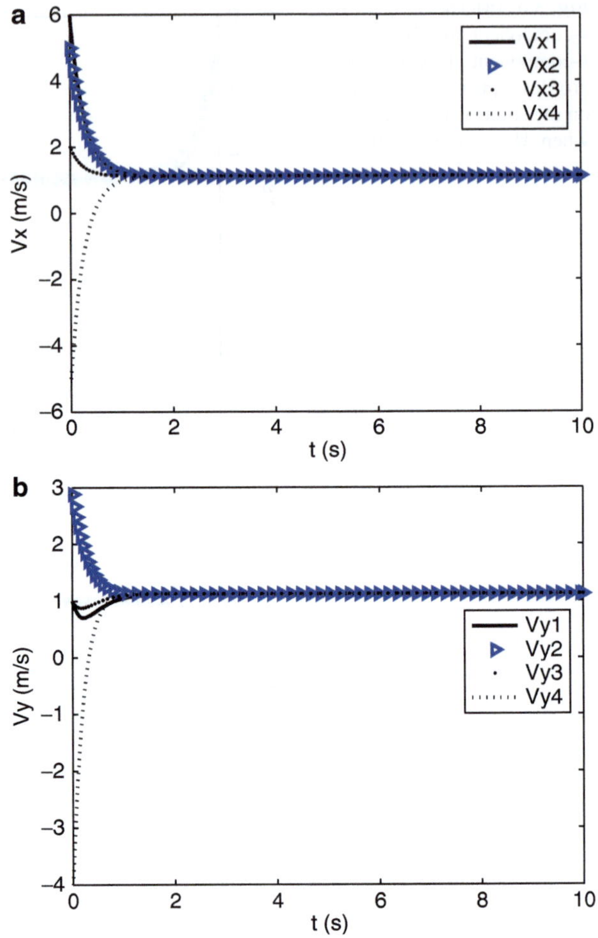

Fig. 5.7 **(a)** The x-component and **(b)** the y-component of the velocity profiles; optimal design based on the solution of the Riccati equation when $AS = 0$ in Example 1

were also considered in Example 1. The graph describing the network structure is shown in Fig. 5.8. The numerical parameters selected for this network of multi-agent systems are $A^1 = \begin{pmatrix} 1 & 1 \\ 0.5 & 1 \end{pmatrix}$, $A^2 = \begin{pmatrix} -2 & 1 \\ 3 & 5 \end{pmatrix}$, $A^3 = \begin{pmatrix} 0 & 3 \\ 4 & -1 \end{pmatrix}$, $A^4 = \begin{pmatrix} 4 & -1 \\ 2 & 1 \end{pmatrix}$, $A^5 = \begin{pmatrix} 2 & -1 \\ -1 & 1 \end{pmatrix}$, and $A^6 = \begin{pmatrix} 6 & 0 \\ 0 & 1 \end{pmatrix}$. The matrices B^i, c^i, R^i, and Q^{ij} are selected as in the previous example. The initial conditions of the velocity vectors are selected as $X^1(0) = [6 \ 1]^T$, $X^2(0) = [8 \ 3]^T$, $X^3(0) = [2 \ 7]^T$, $X^4(0) = [-5 \ -4]^T$, $X^5(0) = [12 \ -6]^T$, and $X^6(0) = [-1 \ -2]^T$. The simulations are obtained by applying the controllers that are designed based on the Riccati equation solutions and the solutions to a set of *LMI*s. In Fig. 5.9a, the x-component and in Fig. 5.9b, the y-component of the velocity profiles of the six-agent team are shown for the above settings in the latter approach. For the sake of comparison in Fig. 5.10a, b the former

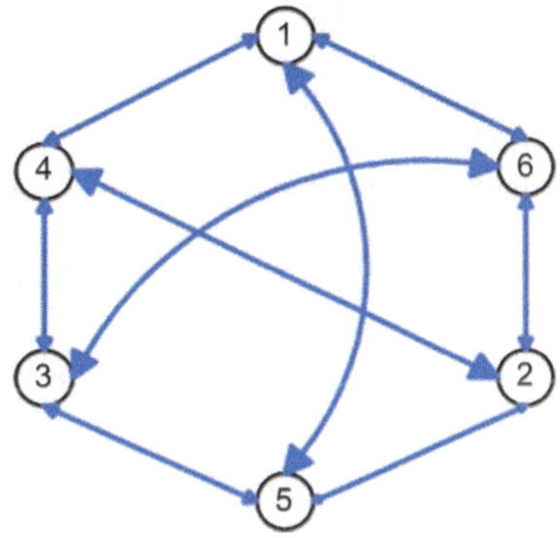

Fig. 5.8 Graph describing the topology of a network of multi-agent systems

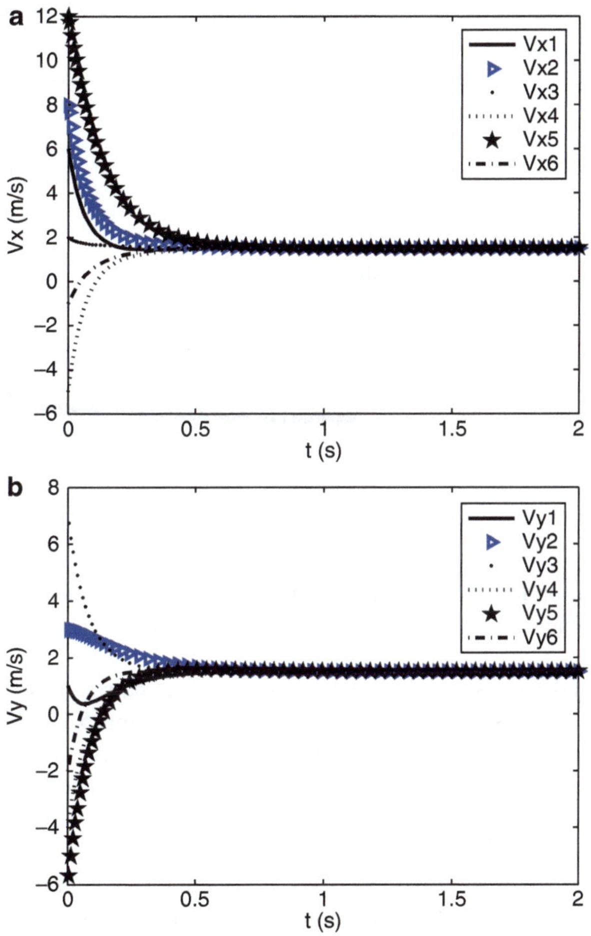

Fig. 5.9 **(a)** The x-component and **(b)** the y-component of the velocity profiles corresponding to an optimal control design based on the solution of the *LMI*s when $AS \neq 0$ in Example 2

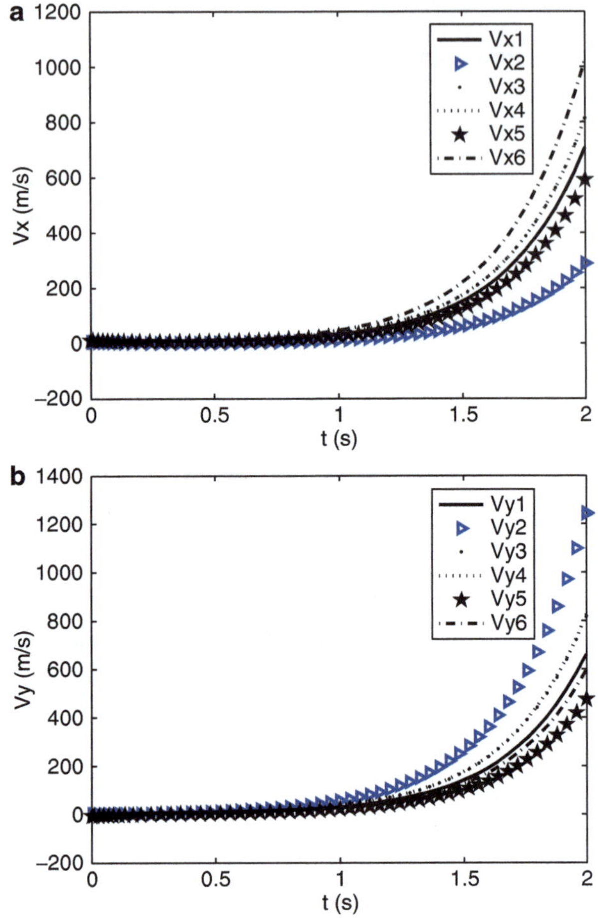

Fig. 5.10 **(a)** The x-component and **(b)** the y-component of the velocity profiles corresponding to an optimal control design based on the solution of the Riccati equation when $AS \neq 0$ in Example 2

strategy is used under the above configuration. It can be seen that the closed-loop system is unstable since the open-loop system matrix A does not satisfy the property $AS = 0$.

In the second case scenario we have selected a matrix A such that $AS = 0$. The corresponding results are presented in Figs. 5.11 and 5.12. To guarantee the condition $AS = 0$, the interaction coefficients $\mathscr{F}^{ij}$ are selected to be zero and the system matrices of all the agents are selected to be the same and given by $A^i = \begin{pmatrix} 1 & -1 \\ -5 & 5 \end{pmatrix}$, $\forall i$. The other simulation parameters are selected to be the same as in the previous scenario. The simulation results are obtained by applying the controllers that are designed based on the Riccati equation solutions and the solutions to a set of *LMI*s. Figure 5.11 corresponds to the latter approach and Fig. 5.12 corresponds to the former approach.

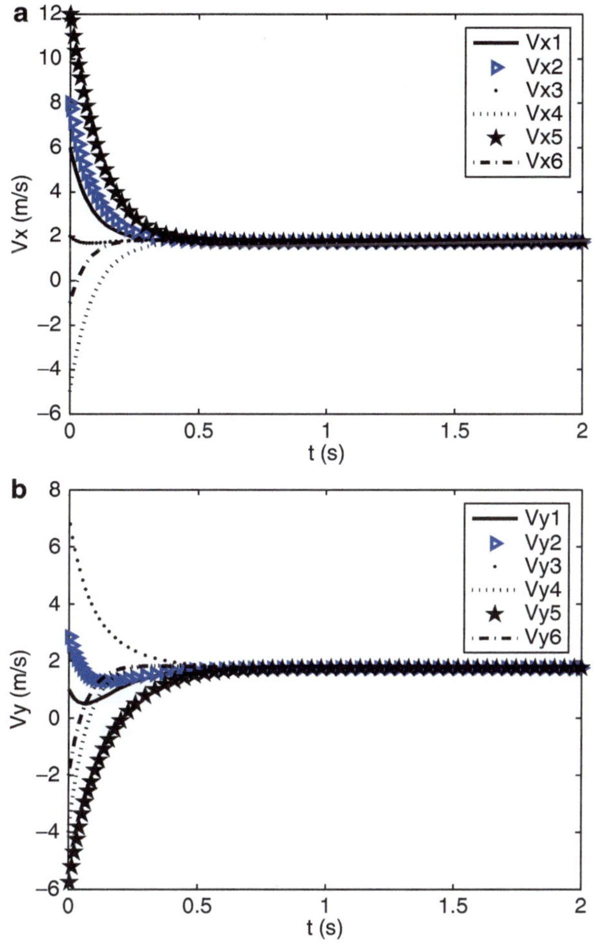

Fig. 5.11 **(a)** The x-component and **(b)** the y-component of the velocity profiles corresponding to an optimal control design based on the solution of the *LMI*s when $AS = 0$ in Example 2

As can be seen from Figs. 5.11 and 5.12 both control methodologies in this case ($AS = 0$) yield an asymptotically stable closed-loop system. For comparison between the two control approaches we have evaluated a performance index corresponding to each method. The numerical values that are obtained for the performance index (5.40) are given in Table 5.2. It can be seen that the cost expensed for the *LMI* approach is indeed higher than the one used by the Riccati equation-based approach when $AS = 0$. Obviously, for the scenario when $AS \neq 0$, the *LMI* approach results in a lower cost given that the Riccati equation-based approach yields an unstable system. On the other hand, the controller provided by the *LMI* approach indeed reaches consensus faster when compared to the controller corresponding to the Riccati equation solution (0.6 s vs. 1.06 s in case of $AS = 0$). The associated numerical results are shown in Table 5.3.

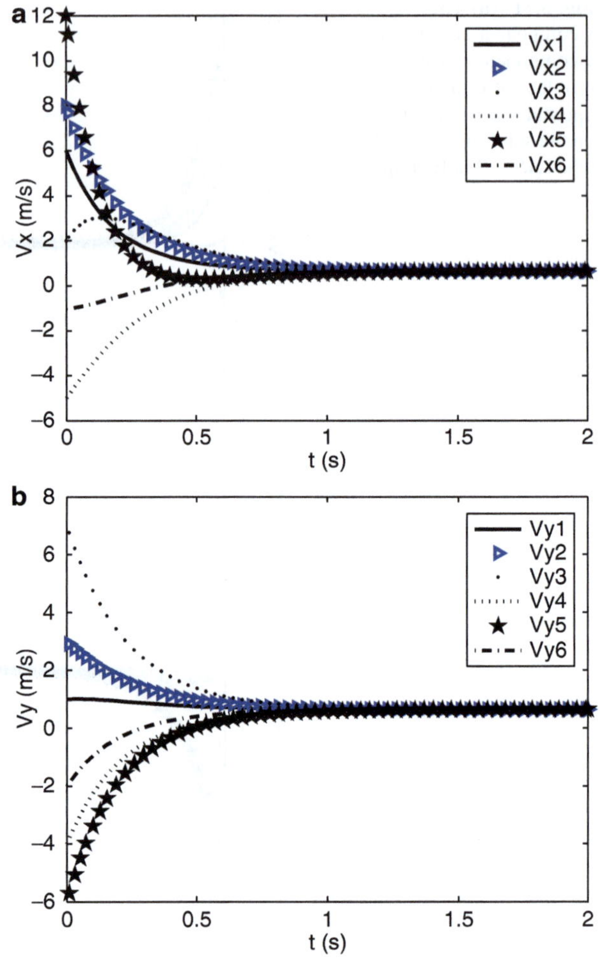

Fig. 5.12 (**a**) The x-component and (**b**) the y-component of the velocity profiles corresponding to an optimal control design based on the solution of the Riccati equation when $AS = 0$ in Example 2

Table 5.2 A comparative evaluation of the performance index corresponding to the two control design strategies for the cost function (5.40) for $T = 2$ s in Example 2

	Performance index (5.40)	
Control scheme	$AS = 0$	$AS \neq 0$
Riccati equation-based	1,079.5	1.0407e+007
LMI-based	2,098.5	2,012.1

Table 5.3 A comparative study of the consensus-reaching time corresponding to the two control design strategies in Example 2

	Consensus time	
Control scheme	$AS = 0$	$AS \neq 0$
Riccati equation-based	1.06 s	–
LMI-based	0.6 s	0.8 s

5.4 Conclusions

In this chapter, the *LMI* formulation is utilized to solve the consensus-seeking problem using game theoretic-based and the optimal control schemes.

First, a novel design-based approach is proposed in order to address the consensus control problem by using a single team cost function within a game theoretic framework. The advantage of minimizing a cost function that describes the total performance of the team is that it can provide a better insight into the performance of the entire team when compared to individual agent performance indices. However, the main disadvantage of this formulation is that it clearly requires the availability of full information set for control design purposes. Our proposed solution alleviates this drawback by imposing and taking into account the information structure of the team by using an *LMI* formulation. It is shown that if the cost function describing the total performance index is minimized, a lower team cost as well as lower individual costs may be achieved. A comparative study is performed between the cooperative game theory strategy and the semi-decentralized optimal control strategy that was introduced in Chap. 3. This comparison reveals that the former approach results in lower individual as well as team cost values as predicted. Moreover, the cooperative game theoretic approach results in a global optimal solution that takes into account the imposed communications constraints.

In the second part of this chapter, an optimal control design strategy that is based on the introduced state decomposition is proposed to guarantee consensus achievement in a network of multi-agent systems. It is shown that the *LMI* optimization provides more flexibility when compared to the method that is based on the solution of the Riccati equation. In other words, the approach based on the solution of the Riccati equation in general fails to provide a solution for a stable consensus protocol. Therefore, the *LMI* formulation is used to solve the corresponding optimization problem and to simultaneously address the consensus achievement constraint. Moreover, by using the *LMI* formulation a controller specific structure based on the neighboring sets can be imposed as an additional *LMI* constraint. Therefore, in the individual control design the only required information will be what has been received from the corresponding neighbors in the controller's neighboring set.

The solutions obtained in this chapter incorporate all the imposed constraints and suggest a global optimal (suboptimal) solution. Also, the proposed formulations provide a single index for describing and analyzing the total performance of the team. These frameworks have sufficient flexibility to accommodate additional constraints and design criteria in the proposed methodologies and solutions. Moreover, since both the game theory and the optimal control are multi-objective frameworks, therefore with the help of the *LMI* formulation the proposed methods have the advantage of being capable of addressing additional specifications, such as limited control input availability (control saturation), specific control structure, and consensus achievement constraints.

Chapter 6
Conclusions and Future Work

6.1 Conclusions

The main focus and emphasis of this book has been on the problem of coordination and collaboration in a network of multi-agent systems. The main objectives pursued have been on addressing the question of how to design a control law to simultaneously satisfy several goals in a team of multi-agent systems, such as requiring stability of the team consensus subject to availability of partial team information, presence of faults, and network communication or topological switching and changes. The problem of team collaboration has played a key role in the development of our proposed methodologies. The main goals of this book are to develop innovative and novel concepts, techniques, and solutions to meet the stringent requirements that are envisaged for the network of unmanned vehicles that are to be employed in different challenging applications.

We have solved the problem of team cooperation for the leaderless and the modified leader–follower structures by using the optimal control theory technique based on the solution of *HJB* equations. The consensus problem is solved for a team of multi-agent systems having a general linear dynamical model or a point-mass model characteristics. We have also introduced interaction terms in the dynamical representation of the multi-agent systems. Stability of the team was guaranteed by using modified consensus algorithms that are achieved by minimizing a set of individual cost functions. In another approach, for obtaining an optimal consensus algorithm, the idea of state decomposition was introduced to simplify the consensus-seeking problem into a stabilization problem. In yet another methodology, the game theory was used to formulate the consensus-seeking problem in a *more* cooperative framework. For this purpose a team cost function was defined and a min–max problem was solved to obtain a cooperative optimal solution for the consensus-seeking problem. It was shown that the results obtained by this approach yield lower cost values when compared to the values that are obtained by using the optimal control techniquc. In both the game theoretic approach and the optimal

E. Semsar-Kazerooni and K. Khorasani, *Team Cooperation in a Network of Multi-Vehicle Unmanned Systems: Synthesis of Consensus Algorithms*, DOI 10.1007/978-1-4614-5073-3_6,

control approach developed based on state decomposition, linear matrix inequalities are used to impose both the decentralized nature of the problem as well as the consensus-seeking constraint on the designed controllers.

We have analyzed the performance and robustness of the designed cooperative controllers in the presence of actuator anomalies for three types of faults. It was shown that depending on the fault type, the steady state error of the agents output may be zero, bounded, or time-varying. The steady state behavior of the team members was discussed and the final value to which each agent converges was predicted for all the three types of faults. The adaptability of the team members to these unanticipated situations and circumstances was also evaluated and verified. Moreover, the assumption of having a fixed and undirected network topology was relaxed to reflect the applicability of our proposed solutions to more realistic and practical circumstances. Specifically, we have considered the possibility of switching topology for the team and assumed that the links among the agents can be bidirectional and weighted as well as time-varying. Even the leader assignment was assumed to be time-varying. It was shown that if the team system matrix corresponding to the error dynamics is designed appropriately, a common Lyapunov function can be found for the team. Therefore, the stability and consensus achievement of the network with a switching structure and switching leader assignment can still be guaranteed. For this purpose, additional constraints should be imposed on the optimal controller coefficients that are designed initially for the fixed network topology. It was shown that by introducing additional criteria, the desirable performance specifications of the team can still be ensured and guaranteed. As a demonstration of such a criterion, the notion of the performance–control effort trade-off was introduced and formally analyzed in detail.

Finally, the team cooperation problem was formulated in a framework which is sufficiently broad and flexible in order to address a wide range of issues and considerations that arise vis-a-vis restricted information exchange structures and agents' dynamics. This work has provided innovative and novel advances to the existing literature on cooperative control by attempting to address more realistic and challenging problems in this domain.

6.2 Further Extensions

Some of the possible extensions of this present work are as follows:

- Generalization of the proposed methodologies to heterogenous types of multi-agent systems and investigation of the consensus protocols for agents having nonlinear dynamics.
- Design of compensating controllers (recovery strategies) that can avoid deterioration of the team performance in the presence of members' faults.
- Incorporation of the stochastic actuator faults into the proposed framework.

- Introduction of a quantified index to measure the effects of decentralization of information on the increase of the proposed team cost. In other words, to quantify the effects of connectedness of the information graph on the performance of the team. This quantization criterion may provide an insight into the trade-offs that exist between the availability of information and the team cost for the proposed methods in this book.
- Obtaining a solution that ensures the required stability of a general switching network structure while the restrictions that are imposed on the optimal performance of the controllers are minimized.
- Considering the cooperative control problem for evolving (dynamic) networks where agents are added or removed from the team.
- Integration of parallel estimation techniques with the common consensus protocols. In many applications, the agents need to estimate the required information due to incomplete measurement and missed or partial information. Hence, either an output of the entire group is available and the state vector of the group should be estimated or part of neighbors' state is available and the rest should be estimated. According to this scenario, an estimate of the state of the entire team (or some of the agents) should be obtained. An important issue is to determine the minimum communication that should be available for each agent to be able to estimate the required information and minimize the disagreement dynamics (observability definition).
- Considering the effects of the noisy communication channels on the proposed methodologies.
- Considering the interaction of several teams while some of them act as adversaries in an environment with dynamic obstacles and pop-up threats.

Any one of the above directions of future research if investigated formally and rigorously can significantly improve the current results on team cooperation and specifically on the consensus-seeking problem.

Appendix A
Proofs

In the following sections we provide the proofs for the lemmas and theorems that are presented in Chaps. 3–5.

A.1 Proofs of the Lemmas and Theorems of Chap. 3

Proof of Lemma 3.1. First note that by assuming $V^i = \frac{1}{2}(v^i)^T K^i(t) v^i + \gamma^i$, Λ^i in (3.11) can be simplified to

$$\Lambda^i(t,X^i,u^{i*}) = \sum_{j\in N^i} (v^i - v^j)^T Q^{ij}(v^i - v^j) + (u_l^{i*})^T R^i u_l^{i*} + \frac{\partial V^i}{\partial v^i}\left(u_l^{i*} + \sum_{j\in N^i} \mathscr{F}^{ij} v^j\right)$$
$$= -\frac{\partial V^i}{\partial t}(t,X^i). \tag{A.1}$$

Now by replacing $V^i, \mathscr{F}^{ij}$, and u_l^{i*} according to (3.13), (3.15), and (3.16), respectively, we obtain

$$-(v^i)^T \frac{\dot{K}^i}{2} v^i - \dot{\gamma}^i = \sum_{j\in N^i} (v^i - v^j)^T Q^{ij}(v^i - v^j) - (v^i)^T \frac{K^i}{4}(R^i)^{-1} K^i v^i$$
$$+ (v^i)^T K^i \sum_{j\in N^i} 2(K^i)^{-1} Q^{ij} v^j. \tag{A.2}$$

By equating the corresponding terms in $(v^i)^T v^i$ and v^i, the Riccati equation in (3.17) may be obtained. This implies that the *HJB* equation (3.11) has a solution which satisfies the boundary conditions, and therefore by using the results presented in [9], it provides the optimal strategy (see [9], p. 222). Moreover, the resulting Riccati equation in (3.17) has a solution since it describes the equation

E. Semsar-Kazerooni and K. Khorasani, *Team Cooperation in a Network of Multi-Vehicle Unmanned Systems: Synthesis of Consensus Algorithms*, DOI 10.1007/978-1-4614-5073-3, © Springer Science+Business Media New York 2013

corresponding to the *LQR* problem in a linear system with the pair (A^i, B^i) described in (3.14). Due to controllability of this pair the solution exists. This completes the proof of this lemma. □

Proof of Theorem 3.1. (a) **Consensus Protocol:** Note that in the infinite horizon case, $\dot{K}^i = 0$ and therefore the differential Riccati equation (3.17) will reduce to an algebraic equation given by

$$2|N^i|Q^{ij} - \frac{1}{2}K^i(R^i)^{-1}K^i = 0 \Rightarrow 2|N^i|(K^i)^{-1}Q^{ij} = \frac{1}{2}(R^i)^{-1}K^i, \tag{A.3}$$

and consequently for $i = 1, \ldots, N$, we obtain

$$u^{i*}(v^i, v^j) = u_l^{i*} + u_g^{i*} = -\frac{1}{2}(R^i)^{-1}K^i \left(v^i - \frac{\sum_{j\in N^i} v^j}{|N^i|} \right) = \Gamma^i \left(v^i - \frac{\sum_{j\in N^i} v^j}{|N^i|} \right). \tag{A.4}$$

(b) and (c) By applying the control law (3.18) to the dynamical equations of each agent in (2.20), the closed-loop velocity dynamics of the entire team will be found as $\dot{v} = L_{cl}v$, in which

$$L_{cl} = \begin{bmatrix} \Gamma^1 & \frac{l_{12}}{|N^1|}\Gamma^1 & \cdots & & \frac{l_{1N}}{|N^1|}\Gamma^1 \\ \frac{l_{21}}{|N^2|}\Gamma^2 & \Gamma^2 & & \cdots & \frac{l_{2N}}{|N^2|}\Gamma^2 \\ \vdots & \vdots & & \vdots & \vdots \\ \frac{l_{i1}}{|N^i|}\Gamma^i & \cdots & & \frac{l_{ij}}{|N^i|}\Gamma^i & \cdots \\ \vdots & \vdots & & \vdots & \vdots \\ \frac{l_{N1}}{|N^N|}\Gamma^N & \cdots & & \frac{l_{N(N-1)}}{|N^N|}\Gamma^N & \Gamma^N \end{bmatrix},$$

where $v = [(v^1)^T \ \ldots \ (v^N)^T]^T$ and l_{ij} is the ijth element of the Laplacian matrix L.

For sake of simplicity and without loss of generality, assume that $|N^i| = \bar{N}, \forall i$, and $R^i = R$, $Q^{ij} = Q \ \forall i, j$ are diagonal PD matrices. Hence, $K^i = K^j, \forall i, j$ and therefore $L_{cl} = \frac{1}{\bar{N}} L \otimes \Gamma$, where $\Gamma = \Gamma^i, \forall i$. It is known from the graph theory that matrix L is always positive semi-definite (PSD) and for undirected connected graphs it has a single zero eigenvalue associated with a unit eigenvector $[1 \ 1 \ \ldots \ 1]^T$. Also, using the properties of Riccati equations and given the fact that the pair (A^i, B^i) defined in (3.14) is reachable and (A^i, Ω^i) is observable, where $Q^{ij} = (\Omega^i)^T\Omega^i$, by putting $\dot{K}^i = 0$ in (3.17), we conclude that the solution K^i is positive definite (PD) [58].

This implies that Γ is diagonal and negative definite (ND). Hence, all the eigenvalues of $L \otimes \Gamma$ are negative except for two (or the size of Γ) zero eigenvalues associated with the eigenvectors: $w(L \otimes \Gamma) = \frac{1}{\sqrt{N}}[1 \ 1 \ \ldots \ 1]^T \otimes w(\Gamma)$, i.e., γ_1 and γ_2. Here, $w(\Gamma)$ denotes any eigenvector of matrix Γ, i.e., w_1, w_2. Therefore, due to the symmetry of $L \otimes \Gamma$, it is negative semi-definite (NSD) with distinct eigenvectors [18].

Now assume that the matrix Δ is the similarity transformation matrix consisting of all the eigenvectors of matrix L_{cl} and define the new state vector to be $\bar{v} = \Delta^{-1}v(t)$. The transformed closed-loop system will be $\dot{\bar{v}} = J\bar{v}$, in which J is the Jordan form of L_{cl} which is fully diagonal with the first two (or size of Γ) diagonal elements being zero and the rest being negative. Therefore, the final state of the closed-loop system in the Jordan canonical form will be $\bar{v}(\infty) = [\bar{v}_1(0)\ \bar{v}_2(0)\ 0\ \ldots\ 0]^T$, and consequently in the original coordinates the steady state value of the vector v will be a linear combination of the first two eigenvectors of L_{cl} corresponding to the zero eigenvalues, i.e., $v(\infty) = \gamma_1\bar{v}_1(0) + \gamma_2\bar{v}_2(0)$. Also, since $\bar{v}(0) = \Delta^{-1}v(0)$, we will have $\bar{v}_1(0) = \gamma_1^T v(0)$ and $\bar{v}_2(0) = \gamma_2^T v(0)$. Hence

$$
\begin{aligned}
v(\infty) &= \gamma_1\bar{v}_1(0) + \gamma_2\bar{v}_2(0) = (\gamma_1\gamma_1^T + \gamma_2\gamma_2^T)v(0) \\
&= \frac{1}{\bar{N}}\left(\begin{bmatrix} 1\ 1 & \cdots & 1 \\ \vdots\ \vdots\ \vdots & & \vdots \\ 1\ 1 & \cdots & 1 \end{bmatrix} \otimes (w_1 w_1^T + w_2 w_2^T)\right) v(0) \\
&= [1\ 1\ \ldots\ 1]^T \otimes (w_1 w_1^T + w_2 w_2^T)\mathrm{Avg}(v(0)).
\end{aligned}
$$

In other words, the final state will be a constant vector of the form $[v^c\ v^c\ \ldots\ v^c]^T$ in which v^c is given by (3.20), and this will lead to the consensus achievement. Moreover, if the matrix Γ has the same diagonal entries, $w_1 w_1^T + w_2 w_2^T$ is identity and hence the consensus value is $\mathrm{Avg}(v(0))$, i.e., an average consensus is achieved. □

Proof of Lemma 3.2. We only prove that the leader case and the followers case can similarly be shown. Note that Λ^1 in (3.22) can be simplified by replacing V^1, u_g^{1*}, and u_l^{1*} according to (3.24), (3.25), and (3.27), respectively

$$
\begin{aligned}
\Lambda^1 = &\sum_{j\in N^1} (v^1 - v^j)^T Q^{1j}(v^1 - v^j) + (v^1 - v^d)^T \Gamma (v^1 - v^d) - (v^1)^T \frac{K^1}{4}(R^1)^{-1}K^1 v^1 \\
&- (g^1)^T \frac{(R^1)^{-1}}{2} K^1 v^1 - (g^1)^T \frac{(R^1)^{-1}}{4} g^1 + ((v^1)^T K^1 + (g^1)^T) \sum_{j\in N^1} 2(K^1)^{-1} Q^{1j} v^j \\
= &-(v^1)^T \frac{\dot{K}^1}{2} v^1 - (\dot{g}^1)^T v^1 - \dot{\gamma}^1.
\end{aligned}
$$

Now, by equating the corresponding terms in $(v^1)^T v^1$ and v^1, the equations in (3.29) and (3.30) can be obtained. □

Proof of Theorem 3.2. (a) **Modified Consensus Protocol:** The proof is similar to the proof of part (a) of Theorem 3.1 and therefore is omitted.

(b) **Stability Analysis:** To prove this part of the theorem let us first illustrate the following fact:

Fact A.1 *For any PD matrix B and matrix A satisfying $A^T + A < 0$ ($A^T + A \leqq 0$), the product $C = BA$ is Hurwitz.*

Proof. We have to show that C satisfies the Lyapunov equation for some PD matrices P and Q, i.e., $C^T P + PC = -Q$. Let us take $P = B^{-1}$. Since B is symmetric, we will have $C^T P + PC = A^T B^T B^{-1} + B^{-1} BA = A + A^T < 0$. This clearly shows that the matrix C is Hurwitz. □

Let us now assume that the desired leader command v^d is time-invariant and define the tracking error for each agent as $e^i = v^i - v^d$. The error dynamics for the entire team can be obtained by using the agents' dynamical equations and the input commands for the leader and followers as given by (2.20), (3.32), and (3.31), respectively, to obtain $\dot{e} = L_{cl} e$ with

$$L_{cl} = \begin{bmatrix} \Gamma^1 + \beta^1 & \frac{l_{12}}{|N^1|}\Gamma^1 & \cdots & & \frac{l_{1N}}{|N^1|}\Gamma^1 \\ \frac{l_{21}}{|N^2|}\Gamma^2 & \Gamma^2 & \cdots & & \frac{l_{2N}}{|N^2|}\Gamma^2 \\ \vdots & \vdots & \vdots & & \vdots \\ \frac{l_{i1}}{|N^i|}\Gamma^i & \cdots & \frac{l_{ij}}{|N^i|}\Gamma^i & & \cdots \\ \vdots & \vdots & \vdots & & \vdots \\ \frac{l_{N1}}{|N^N|}\Gamma^N & \cdots & \frac{l_{N(N-1)}}{|N^N|}\Gamma^N & \Gamma^N & \end{bmatrix}, \tag{A.5}$$

where $e = [(e^1)^T \; \ldots \; (e^N)^T]^T$ and the rest of parameters are defined as before.

For the sake of simplicity, let us assume that $R^i = R$, $Q^{ij} = Q \; \forall i, j$. Hence,

$$L_{cl} = -2K^{-1}(L \otimes Q + G) = -2K^{-1}\left(L \otimes Q + \begin{bmatrix} \Gamma & \vdots & 0 \\ \cdots & \cdots & \cdots \\ 0 & \vdots & 0 \end{bmatrix} \right),$$

where $K = \mathrm{Diag}\{K^i, i = 1, \ldots, N\}$ and $G = \begin{bmatrix} \Gamma & \vdots & 0 \\ \cdots & \cdots & \cdots \\ 0 & \vdots & 0 \end{bmatrix}$. Then, $L \otimes Q$ is PSD with two (or the size of Q) zero eigenvalues associated with the eigenvectors: $w(L \otimes Q) = [1 \; 1 \; \ldots \; 1]^T \otimes w(Q)$, in which $w(Q)$ denotes any eigenvector of the matrix Q. On the other hand, G is PSD because Γ is PD. Hence, $L \otimes Q + G$ is PSD. However, it can be verified that the null-spaces of $L \otimes Q$ and G do not have an intersection, i.e., $\{w | w = [w(Q)^T \; w(Q)^T \; \ldots \; w(Q)^T]^T\} \cap \{w | w = [0 \; \bar{w}]^T\} = \{0\}$, and hence $L \otimes Q + G$ is PD. Also, based on the discussions in the proof of Theorem 3.1, the solution of the Riccati equations obtained by setting $\dot{K}^i = 0$ in (3.28) and $\dot{K}^1 = 0$ in (3.29) is PD and symmetric, and hence K is PD and symmetric. Finally, using the Fact A.1, L_{cl} is Hurwitz, and consequently the error dynamics is asymptotically stable. □

Proof of Lemma 3.3. For the leader case, Λ^1 in (3.36) can be simplified by replacing V^1, u_g^{1*}, and u_l^{1*} according to (3.39), (3.40), and (3.42), respectively. By equating the corresponding terms in $(v^1)^T v^1$, $(v^1)^T v^j$, and v^1, (3.44) and (3.45) may be derived. The follower case is shown along the similar lines. In other words, the HJB equations in (3.35)–(3.37) have a solution that satisfies the boundary conditions so that by using the results developed in [9], an optimal strategy is achieved (see [9], p. 222). Moreover, the Riccati equation in (3.43) [or (3.44)] has a solution since it describes the equation corresponding to an LQR problem for a linear system with a reachable pair (A^i, B^i). Therefore, the solution to (3.43) [or (3.44)] is guaranteed. This, therefore completes the proof of this lemma. □

Proof of Theorem 3.3. (a) Let us first start with the followers case. The leader case may be proved in a similar manner. Note that in the infinite horizon case solutions are achieved by equating $\dot{K}^i = 0$, $\dot{K}^1 = 0$, and $\dot{g}^1 = 0$ in equations (3.43), (3.44), and (3.45), respectively. Consequently, the differential Riccati equation (3.43) reduces to the following algebraic equation:

$$2|N^i|Q^{ij} - \frac{1}{2}K^iB^i(R^i)^{-1}(B^i)^TK^i + (A^i)^TK^i + K^iA^i = 0$$

$$\Rightarrow \frac{1}{2}(R^i)^{-1}(B^i)^TK^i = 2|N^i|(K^iB^i)^{-1}Q^{ij} + (K^iB^i)^{-1}((A^i)^TK^i + K^iA^i),$$

which after some algebraic manipulations yields

$$u^{i*} = -2|N^i|(K^iB^i)^{-1}Q^{ij}\left(v^i - \frac{\sum_{j\in N^i} v^j}{|N^i|}\right) - (K^iB^i)^{-1}((A^i)^TK^i + K^iA^i)v^i.$$

(b) Let us assume that the desired leader command v^d is time-invariant and the tracking error for each agent is given by $e^i = v^i - v^d$. Given the definition of e, the error dynamics is now derived by using the gents' dynamical equations, the input command for the leader, and the input command for the followers as given by (2.21), (3.47), and (3.46), respectively, according to

$$\dot{e}^i = B^i\Gamma^i\left(e^i - \frac{\sum_{j\in N^i} e^j}{|N^i|}\right) - (K^i)^{-1}(A^i)^TK^i(e^i + v^d), \quad i = 2,\ldots,N,$$

$$\dot{e}^1 = B^1\Gamma^1\left(e^1 - \frac{\sum_{j\in N^1} e^j}{|N^1|}\right) + B^1\alpha^1e^1 - (K^1)^{-1}(A^1)^T(g^1 + K^1e^1 + K^1v^d). \tag{A.6}$$

Hence, the error dynamics for the entire team can be written as $\dot{e} = L_{cl}e + f(v^d, g^1)$, where $e = [(e^1)^T \ \ldots \ (e^N)^T]^T$, L_{cl} is given by

$$L_{cl} = -K^{-1}(2L \otimes Q^{ij} + A^TK + 2G),$$

and the matrices K, A, G are defined in the statement of Theorem 3.3. Moreover, $f(v^d, g^1)$ is the part of the error dynamics in (A.6) that is a function of v^d and g^1 which does not affect the stability analysis and is given by

$$f(v^d, g^1) = \begin{bmatrix} -(K^1)^{-1}(A^1)^T(g^1 + K^1 v^d) \\ \vdots \\ -(K^i)^{-1}(A^i)^T K^i v^d \\ \vdots \\ -(K^N)^{-1}(A^N)^T K^N v^d \end{bmatrix}. \tag{A.7}$$

Stability of this matrix can be guaranteed only if matrices Q^{ij}, R^i, and Γ are selected properly such that the matrix L_{cl} is Hurwitz. Intuitively, in order for the followers to achieve a good tracking of the desired output, Q^{ij} and Γ should be selected sufficiently large when compared to R^i. However, in order to obtain a formal solution, one method for finding suitable values for Q^{ij}, R^i, and Γ is to use the LMI technique to ensure stability of the closed-loop system. Here, we assume that Q^{ij}, $\forall i, j$ and R^i, $i = 2, \ldots, N$ are predefined, and therefore we try to find R^1 and Γ such that L_{cl} is Hurwitz. For this purpose, let the following *LMI* in variables Γ, K^1 be satisfied:

$$\Upsilon + \Upsilon^T > 0, \ \ \Upsilon = -KL_{cl} = 2L \otimes Q^{ij} + A^T K + 2G.$$

As discussed previously the pair (A^i, B^i) that is used in (2.21) is reachable, and $Q^{ij} = (\Omega^i)^T \Omega^i > 0$ may always be selected such that the matrix Ω^i is full rank, and hence the pair (A^i, Ω^i) is observable. Therefore, the solution to the Riccati equations obtained by setting $\dot{K}^i = 0$ in (3.43) and $\dot{K}^1 = 0$ in (3.44) will be *PD*, leading to K to also be *PD*. Now by invoking the Fact A.1, one may conclude that $L_{cl} = -K^{-1}\Upsilon$ is Hurwitz. On the other hand, K^1 has to satisfy the Riccati equation (3.44) and therefore this equation should be added as a constraint. Hence, the following set of *LMI*s may be considered in which the unknowns to be determined are the elements Γ, K^1, and $Z^1 = -\frac{1}{2}K^1 B^1 (R^1)^{-1}(B^1)^T K^1$, governed by the expressions

$$\begin{cases} \Upsilon + \Upsilon^T > 0, \ \ \Upsilon = 2L \otimes Q^{ij} + A^T K + 2G, \\ (A^1)^T K^1 + K^1 A^1 + Z^1 + 2(|N^1| Q^{1j} + \Gamma) = 0, \\ \Gamma > 0, \ \ K^1 > 0, \ \ Z^1 < 0, \end{cases}$$

and from which Γ, K^1, and Z^1, and hence $R^1 = -\frac{1}{2}(B^1)^T K^1 (Z^1)^{-1} K^1 B^1$ may be calculated.

Note that the matrix Q^{1j} is assumed to be predefined and is set equal to $Q^{ij}, i = 2, \ldots, N$. However, Q^{1j} can also be selected and designed formally by adding it to the set of LMI parameters. □

Proof of Theorem 3.4. (a) The proof for this part is similar to the proof of part (a) of Theorem 3.3.
(b), (c) We first note that for achieving consensus one should determine the closed-loop dynamics corresponding to the team velocity. Using the results of part (a) and the combined control law given by (3.53), the dynamics of each agent's velocity will be governed by

$$\dot{v^i} = A^i v^i - 2(K^i)^{-1}|N^i|Q^{ij}\left(v^i - \frac{\sum_{j\in N^i} v^j}{|N^i|}\right) - (K^i)^{-1}(K^i A^i + (A^i)^T K^i)v^i$$

$$= -2(K^i)^{-1}|N^i|Q^{ij}\left(v^i - \frac{\sum_{j\in N^i} v^j}{|N^i|}\right) - (K^i)^{-1}(A^i)^T K^i v^i.$$

Consequently, the closed-loop system team matrix becomes

$$L_{cl} = -K^{-1}(2L \otimes Q^{ij} + A^T K),$$

where $\dot{v} = L_{cl}v$, $v = [(v^1)^T \ (v^2)^T \ \ldots \ (v^N)^T]^T$, $K = \text{Diag}\{K^i, i = 1,\ldots,N\}$, and $A = \text{Diag}\{A^i, i = 1,\ldots,N\}$. Now in order to have consensus, L_{cl} should be stable and moreover we should have

$$L_{cl}w = 0,\ w = [(v^c)^T \ (v^c)^T \ \ldots \ (v^c)^T]^T, \tag{A.8}$$

which implies that $2L \otimes Q^{ij}w + A^T K w = 0$. The first term is always zero due to the properties of the Laplacian matrix, i.e., $2L \otimes Q^{ij}w = 0$, and therefore we should have

$$A^T K w = 0 \Rightarrow\ (A^i)^T K^i v^c = 0. \tag{A.9}$$

Due to the nonsingularity of K^i, the above expression implies that the matrix A^i should be rank deficient, i.e., $|(A^i)^T| = 0$ and v^c should be the eigenvector of $(A^i)^T K^i$ corresponding to the zero eigenvalue. This implies that v^c should be in the null space of $(A^i)^T K^i$.

Moreover, in order to have a stable consensus, $L_{cl} = -K^{-1}(2L \otimes Q^{ij} + A^T K)$ should be stable, i.e. all its eigenvalues should be negative except the zeros corresponding to the eigenvector w. For this to hold and since K^{-1} is PD it suffices to have $(2L \otimes Q^{ij} + A^T K) + (2L \otimes Q^{ij} + A^T K)^T \geq 0$ (see Fact A.1). Equivalently, we should have $2L \otimes Q^{ij} + (2L \otimes Q^{ij})^T + A^T K + KA \geq 0$. However, $2L \otimes Q^{ij} + (2L \otimes Q^{ij})^T \geq 0$ holds by the definition of the Laplacian matrix and connectivity of the graph. Therefore, it is sufficient to have $A^T K + KA \geq 0$ or $(A^i)^T K^i + K^i A^i \geq 0$. □

A.2 Proofs of the Lemmas and Theorems of Chap. 4

Proof of Lemma 4.1. (a), (b) Here, we only discuss the followers' failure. It should be noted that with minor modifications the proof presented here can accommodate *LOE* faults in the leader as well.

Based on the discussions given in the proof of Theorem 3.2, the error dynamics closed-loop matrix L_{cl} can be written as

$$L_{cl} = -2K^{-1}(L \otimes Q + G) = -2K^{-1}\left(L \otimes Q + \begin{bmatrix} \Gamma & \vdots & 0 \\ \cdots & \cdots & \cdots \\ 0 & \vdots & 0 \end{bmatrix}\right). \tag{A.10}$$

When an *LOE* fault is injected at some followers the closed-loop dynamics of the system is governed by (4.2), with the closed-loop matrix given by

$$L_f = \begin{bmatrix} L_{11} & \vdots\, L_{12} \\ \cdots\cdots & \cdots\cdots \\ \alpha L_{21} & \vdots\, \alpha L_{22} \end{bmatrix} = I_f L_{cl} = -2I_f K^{-1}(L \otimes Q + G)$$

$$= -2I_f \begin{bmatrix} (K_{11}^{-1})_{m(N-q)\times m(N-q)} & \vdots\, 0_{m(N-q)\times mq} \\ \cdots\cdots & \cdots\cdots \\ 0_{mq\times m(N-q)} & \vdots\, (K_{22}^{-1})_{mq\times mq} \end{bmatrix} \times \begin{bmatrix} (\bar{L}_{11} + G_{11}) & \vdots\, (\bar{L}_{12} + G_{12}) \\ \cdots\cdots & \cdots\cdots \\ (\bar{L}_{21} + G_{21}) & \vdots\, (\bar{L}_{22} + G_{22}) \end{bmatrix}, \tag{A.11}$$

where $K_{ij}^{-1}, \bar{L}_{ij}, G_{ij}$ denote the partitions of matrices K^{-1}, $L \otimes Q$, and G, respectively, which correspond to the faulty and healthy agents' dynamics. Moreover, I_f is defined as

$$I_f = \begin{bmatrix} I_{m(N-q)\times m(N-q)} & \vdots\, 0_{m(N-q)\times mq} \\ \cdots\cdots & \cdots\cdots \\ 0_{mq\times m(N-q)} & \vdots\, \alpha I_{mq\times mq} \end{bmatrix}. \tag{A.12}$$

The above matrix is PD and diagonal for $0 < \alpha \leq 1$, and therefore $I_f K^{-1}$ is also PD. In addition, $L \otimes Q + G$ can be shown to be a PD matrix. Hence, by invoking the Fact A.1, it can be seen that L_f is asymptotically stable, and consequently the error asymptotically approaches to zero. This implies that the consensus achievement is guaranteed and all the agents' output will converge to the command provided by the leader. □

Proof of Lemma 4.2. (a) As discussed previously, the closed-loop error dynamics matrix L_{cl} can be written as

$$L_{cl} = -2K^{-1}(L \otimes Q + G).$$

In case of faults in some followers the closed-loop dynamics will be governed by (4.3), where the closed-loop matrix can be written as

$$L_f = \begin{bmatrix} L_{11} & \vdots\, L_{12} \\ \cdots\cdots & \cdots\cdots \\ 0_{mq\times m(N-q)} & \vdots\, 0_{mq\times mq} \end{bmatrix} = I_f L_{cl} = -2I_f K^{-1}(L \otimes Q + G)$$

$$= -2I_f \begin{bmatrix} (K_{11}^{-1})_{m(N-q)\times m(N-q)} & \vdots\, 0_{m(N-q)\times mq} \\ \cdots\cdots & \cdots\cdots \\ 0_{mq\times m(N-q)} & \vdots\, (K_{22}^{-1})_{mq\times mq} \end{bmatrix} \times \begin{bmatrix} (\bar{L}_{11}+G_{11}) & \vdots\, (\bar{L}_{12}+G_{12}) \\ \cdots\cdots & \cdots\cdots \\ (\bar{L}_{21}+G_{21}) & \vdots\, (\bar{L}_{22}+G_{22}) \end{bmatrix}, \tag{A.13}$$

where $K_{ij}^{-1}, \bar{L}_{ij}, G_{ij}$ denote the corresponding partitions of the matrices K^{-1}, $L \otimes Q$, and G, respectively, which correspond to the faulty and healthy agents' dynamics. Moreover, I_f is defined as follows:

$$I_f = \begin{bmatrix} I_{m(N-q)\times m(N-q)} & \vdots\, 0_{m(N-q)\times mq} \\ \cdots\cdots & \cdots\cdots \\ 0_{mq\times m(N-q)} & \vdots\, 0_{mq\times mq} \end{bmatrix}. \tag{A.14}$$

The above matrix is *PSD* and so $I_f K^{-1}$ is *PSD*. Also, $L \otimes Q + G$ is shown to be a *PD* matrix. Hence, by invoking the Fact A.1, it can be seen that L_f will be stable and hence the error remains bounded.

(b) It can be readily verified that the eigenvalues of L_f, consist of eigenvalues of matrix L_{11} and mq zeros. Using (A.13), L_{11} can be written as $L_{11} = -2K_{11}^{-1}(\bar{L}_{11} + G_{11})$ and due to the positive definiteness of $L \otimes Q + G$ and K^{-1}, $\bar{L}_{11} + G_{11}$ and K_{11}^{-1} are both PD, and therefore L_{11} is Hurwitz (see Fact A.1). Hence, its eigenvalues are all negative, and therefore the only zero eigenvalues of L_f are due to the zero rows of the matrix L_f. Moreover, the eigenvectors of L_f correspond to zero eigenvalues, i.e., $w(L_f)$ are as follows:

$$w(L_f) \in \{w | w \in \mathrm{Null}\{[L_{11} \,\vdots\, L_{12}]\}\}, \tag{A.15}$$

which provides mq distinct eigenvectors corresponding to zero eigenvalues. This is due to the fact that the rows of the matrix $[L_{11} \,\vdots\, L_{12}]$ are linearly independent (L_{cl} is full rank) and hence the rank of Null$\{[L_{11} \vdots L_{12}]\}$ is mq, which can provide mq independent eigenvectors for L_f corresponding to zero eigenvalues. This results in diagonal Jordan blocks corresponding to zero eigenvalues, and hence the final value of the error vector will be a linear combination of these eigenvectors and therefore is in the Null$\{[L_{11} \vdots L_{12}]\}$, i.e.,

$$\begin{aligned}
& e_{ss} \in \{w | w = [w_1^T \ \ w_2^T]^T \in \text{Null}\{[L_{11} \vdots L_{12}]\}\} \\
& \Rightarrow \left[L_{11} \vdots L_{12} \right] e_{ss} = L_{11} e_{wss} + L_{12} e_{fss} = 0 \\
& \Rightarrow e_{wss} = -L_{11}^{-1} L_{12} e_{fss} = -L_{11}^{-1} L_{12} e^f \Rightarrow e_{ss} = \begin{bmatrix} -L_{11}^{-1} L_{12} \\ \dots\dots\dots \\ I_{mq \times mq} \end{bmatrix} e^f. \qquad \text{(A.16)}
\end{aligned}$$

□

Proof of Lemma 4.3. (a), (b) In case of a leader fault, the closed-loop dynamics will be governed by (4.4) and the closed-loop matrix can be written as

$$\begin{aligned}
L_f &= \begin{bmatrix} 0_{m \times m} \vdots 0_{m \times (N-1)m} \\ \dots\dots\dots\dots\dots\dots \\ L_{21} \quad \vdots L_{22} \end{bmatrix} = I_f L_{cl} = -2 I_f K^{-1} (L \otimes Q + G) = -2K^{-1} ((I_{ff} L) \otimes Q) \\
&= \frac{1}{|N^i|} (I_{ff} L) \otimes \Gamma^i = \begin{bmatrix} 0_{m \times m} & \vdots \ 0_{m \times (N-1)m} \\ \dots\dots\dots\dots & \dots\dots\dots\dots \\ (L_{21})_{(N-1)m \times m} & \vdots \ (L_{22})_{(N-1)m \times (N-1)m} \end{bmatrix}, \qquad \text{(A.17)}
\end{aligned}$$

where Γ^i is defined as in (3.33) and I_f and I_{ff} are defined as follows:

$$I_f = \begin{bmatrix} 0_{m \times m} & \vdots \ 0_{m \times m(N-1)} \\ \dots\dots\dots & \dots\dots\dots \\ 0_{m(N-1) \times m} & \vdots \ I_{m(N-1) \times m(N-1)} \end{bmatrix}, \ I_{ff} = \begin{bmatrix} 0 & \vdots \ 0_{1 \times (N-1)} \\ \dots\dots\dots & \dots\dots\dots \\ 0_{(N-1) \times 1} & \vdots \ I_{(N-1) \times (N-1)} \end{bmatrix}. \qquad \text{(A.18)}$$

It can be shown that multiplying the Laplacian matrix by any matrix of kind I_{ff} will not change the spectrum of the Laplacian matrix [41]. Namely, it will have the same number of zero eigenvalues (the same inertia). Moreover, it is known from graph theory that L is always *PSD* and for undirected connected graphs it has a single zero eigenvalue associated with a unit eigenvector, i.e.: $[1 \ 1 \ \dots \ 1]^T$. This implies that $I_{ff} L$ has one zero eigenvalue and $N-1$ positive eigenvalues. Also, without loss of generality, assume that both Q^{ij} and R^i are diagonal. On the other hand, using properties of the Riccati equations and due to the observability and reachability conditions, the solution of the Riccati equation, K^i, obtained by putting $\dot{K}^i = 0$ in (3.28) will be diagonal and PD (see [58]). This implies that Γ^i is diagonal and negative definite.

Using properties of the Kronecker product, all the eigenvalues of $(I_{ff}L)\otimes\Gamma^i$ are negative except for two (or the size of Γ^i) zero eigenvalues associated with distinct eigenvectors: $\frac{1}{\sqrt{N}}[1\ 1\ \ldots\ 1]^T \otimes w(\Gamma^i)$. Here $w(\Gamma^i)$ denotes any eigenvector of the matrix Γ^i. Hence, $(I_{ff}L)\otimes\Gamma^i$ is stable. Now, using similar discussion as in the proof of Lemma 4.2, e_{ss} will be in the Null$\{[L_{21} \vdots L_{22}]\}$ which has rank m (here 2). It can be seen that any element in this space is of the form $[(e^f)^T\ (e^f)^T\ \ldots\ (e^f)^T]^T$, i.e., $e_{ss} = [(e^f)^T\ (e^f)^T\ \ldots\ (e^f)^T]^T$. □

Proof of Theorem 4.2. **Leader Adaptability:** Without loss of generality, assume that $m = 1$ or e_f (e^f) is a scalar and $q = 1$, i.e. single-input single-output subsystems. In order to show that $e^1 \bullet e^f > 0$, it is sufficient to show that $[e^f\ 0\ \ldots\ 0]_{1\times(N-1)}e_{wss} > 0$, where e_{wss} is the steady state value of the vector e_w. In the following we try to simplify this condition as much as possible to finally verify its correctness. Using part (b) of Lemma 4.2, this condition reduces to

$$\begin{aligned}[e^f\ 0\ \ldots\ 0]_{1\times(N-1)}e_{wss} > 0 &\leftrightarrow -(e^f)^2[1\ 0\ \ldots\ 0]L_{11}^{-1}L_{12} > 0\\ &\leftrightarrow [1\ 0\ \ldots\ 0]L_{11}^{-1}L_{12} < 0. \end{aligned} \tag{A.19}$$

Also, based on the definition of L_{cl} given in (4.1), I_f given in (A.14) and L_f given in (A.13), we have

$$\begin{aligned} L_f[1\ 1\ \ldots\ 1]^T_{1\times N} &= I_f L_{cl}[1\ 1\ \ldots\ 1]^T_{1\times N} = [\beta^1\ 0\ \ldots\ 0]^T_{1\times N}\\ &\leftrightarrow L_{11}[1\ 1\ \ldots\ 1]^T_{1\times(N-1)} + L_{12} = [\beta^1\ 0\ \ldots\ 0]^T_{1\times(N-1)}\\ &\leftrightarrow L_{12} = -L_{11}[1\ 1\ \ldots\ 1]^T + [\beta^1\ 0\ \ldots\ 0]^T. \end{aligned}$$

By replacing the above value for L_{12} in (A.19) and given that $\beta^1 < 0$, one gets

$$[1\ 0\ \ldots\ 0]L_{11}^{-1}L_{12} = [1\ 0\ \ldots\ 0]L_{11}^{-1}[\beta^1\ 0\ \ldots\ 0]^T - 1 < 0 \leftrightarrow L_{11}^{-1}(1,1) > \frac{1}{\beta^1}, \tag{A.20}$$

where $L_{11}^{-1}(i,j)$ is the ijth element of the matrix L_{11}^{-1}. Now, in order to show that (A.20) is satisfied, we can use the properties of the inverse of a matrix to describe this element as

$$L_{11}^{-1}(1,1) = \frac{C_{11}}{\det(L_{11})} = \frac{C_{11}}{L_{11}(1,1)C_{11}+\cdots+L_{11}(1,N-1)C_{1(N-1)}}$$
$$= \frac{C_{11}}{(\Gamma^1+\beta^1)C_{11}+\cdots+L_{11}(1,N-1)C_{1(N-1)}} = \frac{C_{11}}{\beta^1 C_{11}+\det(-2K_{11}^{-1}\bar{L}_{11})},$$

where $L_{11}(i,j)$ is the ijth element of the matrix L_{11} and C_{ij} denotes the ijth cofactor of L_{11}. Also, $K_{ij}^{-1}, \bar{L}_{ij}$ denote the corresponding partitions of the matrices K^{-1}, $L \otimes Q$, respectively, which correspond to the faulty and healthy agents' dynamics, and Γ^1, β^1 are defined in (4.1). Without loss of generality, assume that N, i.e., the total number of agents, is an odd number. Similar reasoning may be used for an even N. Since L_{11} is $(N-1) \times (N-1)$, $N-1$ is even, and both K_{11}^{-1} and $\bar{L}_{11}+G_{11}$ are PD matrices and $N-1$ is even, then the determinant of the matrix $L_{11} = -2K_{11}^{-1}(\bar{L}_{11}+G_{11})$ is a positive number. Hence, $\beta^1 C_{11}+\det(-2K_{11}^{-1}\bar{L}_{11})$ is positive. Similarly, $\det(-2K_{11}^{-1}\bar{L}_{11})$ is also positive, since both K_{11}^{-1} and $\bar{L}_{11}$ are PD matrices. Therefore, $\beta^1 C_{11}+\det(-2K_{11}^{-1}\bar{L}_{11}) > \beta^1 C_{11}$, and hence $\frac{\beta^1 C_{11}}{\beta^1 C_{11}+\det(-2K_{11}^{-1}\bar{L}_{11})} < 1$ or equivalently $\frac{C_{11}}{\beta^1 C_{11}+\det(-2K_{11}^{-1}\bar{L}_{11})} > \frac{1}{\beta^1}$. The correctness of this inequality guarantees that the initial inequality, i.e., $e^1 \bullet e^f > 0$, is true. This completes the proof of this part.

Followers' Adaptability: In order to show that $e^k \bullet e^f > 0$, $k = 2,\ldots,N$, it is sufficient to show that $[0\ \ldots\ (e^f)_k\ \ldots\ 0]e_w > 0$. By using part (b) of Lemma 4.2, this condition reduces to

$$-(e^f)^2[0\ \ldots\ (1)_k\ \ldots\ 0]L_{11}^{-1}L_{12} > 0 \rightarrow [0\ \ldots\ (1)_k\ \ldots\ 0]L_{11}^{-1}L_{12} < 0. \quad \text{(A.21)}$$

Similar to the leader's case we have

$$L_{12} = -L_{11}[1\ 1\ \ldots\ 1]^T + [\beta^1\ 0\ \ldots\ 0]^T.$$

By replacing the above value for L_{12} in (A.21) and given that $\beta^1 < 0$, we obtain

$$[0\ \ldots\ (1)_k\ \ldots\ 0]L_{11}^{-1}L_{12} = [0\ \ldots\ (1)_k\ \ldots\ 0]L_{11}^{-1} \times [\beta^1\ 0\ \ldots\ 0]^T - 1 < 0$$
$$\leftrightarrow L_{11}^{-1}(k,1) > \frac{1}{\beta^1}. \quad \text{(A.22)}$$

By using the properties of the inverse of a matrix we have

$$L_{11}^{-1}(k,1) = \frac{C_{1k}}{\det(L_{11})} = \frac{C_{1k}}{L_{11}(1,1)C_{11}+L_{11}(1,2)C_{12}+\cdots+L_{11}(1,N-1)C_{1(N-1)}}$$
$$= \frac{C_{1k}}{(\Gamma^1+\beta^1)C_{11}+\cdots+L_{11}(1,N-1)C_{1(N-1)}}$$
$$= \frac{C_{1k}}{\beta^1 C_{11}+\det(-2K_{11}^{-1}\bar{L}_{11})} > \frac{1}{\beta^1}, \quad \text{(A.23)}$$

where C_{ij}, K_{11}^{-1}, $\bar{L}_{11}$, Γ^1, β^1 are defined as before. As discussed above in the first part of the proof both $\beta^1 C_{11} + \det(-2K_{11}^{-1}\bar{L}_{11})$ and $\det(-2K_{11}^{-1}\bar{L}_{11})$ are positive. On the other hand, $\beta^1 C_{11}$ is a function of Γ and can take on arbitrary large values regardless of the magnitude of $\det(-2K_{11}^{-1}\bar{L}_{11})$. Therefore, in order to maintain the two conditions $\det(-2K_{11}^{-1}\bar{L}_{11}) > 0$ and $\beta^1 C_{11} + \det(-2K_{11}^{-1}\bar{L}_{11}) > 0$, the term $\beta^1 C_{11}$ has to be positive, and therefore C_{11} should be negative.

Now, in order to prove that (A.23) holds and given that $\beta^1 < 0$, we should show the following inequality:

$$\beta^1 C_{1k} < \beta^1 C_{11} + \det(-2K_{11}^{-1}\bar{L}_{11}).$$

For this to hold it suffices to have $\beta^1 C_{1k} < \beta^1 C_{11}$. Since, $C_{11} < 0$, if we can show that $|C_{1k}| < |C_{11}|$, then (A.23) will be guaranteed. In the following we show this property.

First, note that for any $k > 1$, we have the following property:

$$C_{1k}(L_{11}) = C_{1k}(-2K_{11}^{-1}(\bar{L}_{11} + G_{11})) = C_{1k}(-2K_{11}^{-1}\bar{L}_{11}). \tag{A.24}$$

The last equality holds as a result of the definition of matrix G_{11}. Also, $C_{1k}(-2K_{11}^{-1}\bar{L}_{11}) = (-1)^{1+k}M_{1k}(-2K_{11}^{-1}\bar{L}_{11})$. However, $M_{1k}(-2K_{11}^{-1}\bar{L}_{11})$ is equal to the 2×2 minor of the matrix $\bar{L} = -2K^{-1}(L\otimes Q)$, resulting from the deletion of rows $1,N$ and columns k,N. Similarly $M_{11}(-2K_{11}^{-1}\bar{L}_{11})$ is the 2×2 minor resulting from the deletion of rows and columns $1,N$ of the matrix $\bar{L}$. In other words, $M_{1k}(-2K_{11}^{-1}\bar{L}_{11}) = \det(\bar{L}(1,N|k,N))$ and $M_{11}(-2K_{11}^{-1}\bar{L}_{11}) = \det(\bar{L}(1,N|1,N))$.

Also, $-\bar{L}$ can be considered as the Laplacian matrix of a weighted graph. Therefore, based on the assumption that the information structure is described by a tree-like graph, the result of Theorem 4.1 may be used. For the general proof when the graph consists of some cycles, the general form of the matrix tree theorem presented in [22] should be used. Using Theorem 4.1, one may describe $|\det(\bar{L}(1,N|1,N))|$ and $|\det(\bar{L}(1,N|k,N))|$ as follows:

$$\begin{aligned} |\det(\bar{L}(1,N|1,N))| &= \text{length}(P(v_1,v_N)\cap P(v_1,v_N)) = \text{length}(P(v_1,v_N)),\\ |\det(\bar{L}(1,N|k,N))| &= \text{length}(P(v_1,v_N)\cap P(v_k,v_N)). \end{aligned} \tag{A.25}$$

The second path is the intersection of two paths $P(v_1,v_N)$ and $P(v_k,v_N)$ which is obviously a subset of $P(v_1,v_N)$, and hence its length is smaller. Consequently, $|\det(\bar{L}(1,N|k,N))|$ is smaller than $|\det(\bar{L}(1,N|1,N))|$. This in turn implies that $|C_{1k}| < |C_{11}|$. □

Proof of Lemma 4.4. When an *LIP* fault occurs, the dynamics of the faulty and healthy agents in the team are governed by the following equations:

$$\dot{e}_f = u_c \Rightarrow e_f = u_c t - u_c t_f + e^f,\ \ \dot{e}_w = L_{11}e_w + L_{12}e_f = L_{11}e_w + L_{12}(u_c t - u_c t_f + e^f). \tag{A.26}$$

Therefore, e_w can be obtained as

$$\begin{aligned}
e_w(t) &= e^{L_{11}(t-t_f)}e_w(t_f)+\int_{t_f}^{t} e^{L_{11}(t-\tau)}L_{12}(u_c\tau-u_ct_f+e^f)\mathrm{d}\tau\\
&= e^{L_{11}(t-t_f)}e_w(t_f)-[e^{L_{11}(t-\tau)}L_{11}^{-1}L_{12}(u_c\tau-u_ct_f+e^f)+e^{L_{11}(t-\tau)}L_{11}^{-2}L_{12}u_c]_{t_f}^{t}\\
&= e^{L_{11}(t-t_f)}[e_w(t_f)+L_{11}^{-1}L_{12}e^f-L_{11}^{-2}L_{12}u_c]-[L_{11}^{-1}L_{12}e^f+L_{11}^{-2}L_{12}u_c]\\
&\quad -L_{11}^{-1}L_{12}u_ct+L_{11}^{-1}L_{12}u_ct_f.
\end{aligned} \tag{A.27}$$

Similar to the discussion that is presented in the proof of Lemma 4.2, it can be shown that L_{11} is Hurwitz. Therefore, in the steady state, the effects of the first term in the above expression vanish asymptotically, since $e^{L_{11}(t-t_f)}\to 0$ as $t\longrightarrow\infty$. Therefore, the dominant expression of $e_w(t)$ as $t\to\infty$ is governed by

$$e_w(t)\to -[L_{11}^{-1}L_{12}e^f+L_{11}^{-2}L_{12}u_c]-L_{11}^{-1}L_{12}u_ct+L_{11}^{-1}L_{12}u_ct_f, \tag{A.28}$$

and consequently e_{ss} converges to the value that is given by (4.9). Obviously, the error dynamics is not stable in this case. □

Proof of Lemma 4.5. According to the discussion given in Sect. 4.1.3, the error dynamics for the entire team can be written as $\dot{e}=L_{cl}e+f(v^d,g^1)$, where L_{cl} is defined as

$$L_{cl}=-K^{-1}(2L\otimes Q^{ij}+A^TK+2G)=-K^{-1}\Upsilon, \tag{A.29}$$

and K, A, G, Υ, $f(v^d,g^1)$ are defined as before. When an *LIP* fault occurs in a follower, the closed-loop dynamics is then governed by

$$\dot{e}=\begin{bmatrix}\dot{e}_w\\ \dot{e}_f\end{bmatrix}=\left[\begin{array}{c:c}(L_{11})_{m(N-1)\times m(N-1)} & (L_{12})_{m(N-1)\times m}\\ \hdashline 0_{m\times m(N-1)} & A^f\end{array}\right]e+\begin{bmatrix}f_1(v^d,g^1)\\ A^fv^d+B^fu_c\end{bmatrix}. \tag{A.30}$$

The above dynamical system is stable if both matrices L_{11} and A^f are stable. The latter is true by assumption and the former can be shown as follows. Towards this end, we should note that L_{11} can be written as follows:

$$L_{11}=-K_{11}^{-1}\Upsilon_{11}, \tag{A.31}$$

where $K_{ij}^{-1},\Upsilon_{ij}$ denote the corresponding partitions of the matrices K^{-1}, Υ, respectively, which correspond to the faulty and healthy agents' dynamics. Now, from

Theorem 3.3, we know that $\Upsilon + \Upsilon^T > 0$, and therefore its partition Υ_{11} enjoys the same property, i.e., $\Upsilon_{11} + \Upsilon_{11}^T > 0$. This is due to the fact that $\Upsilon_{11} + \Upsilon_{11}^T$ is a principal minor of $\Upsilon + \Upsilon^T$ and since $\Upsilon + \Upsilon^T$ is *PD*, any of its principal minors is also *PD*. Moreover, K_{11}^{-1} is *PD* for the similar reason. Now, invoking the Fact A.1, we can conclude that L_{11} is stable. Hence, the entire error dynamics is stable. Given that A^f is Hurwitz, $(A^f)^{-1}$ is defined and hence in the steady state we have

$$\begin{aligned} \dot{e}_f &= A^f(e_f + v^d) + B^f u_c = 0 \Rightarrow (e_f)_{ss} = -(A^f)^{-1} B^f u_c - v^d, \\ \dot{e}_w &= L_{11} e_w + L_{12} e_f + f_1(v^d, g^1) \Rightarrow (e_w)_{ss} = L_{11}^{-1}[L_{12}((A^f)^{-1} B^f u_c + v^d) - f_1(v^d, g^1)]. \end{aligned} \tag{A.32}$$

This completes the proof of this lemma. □

Proof of Lemma 4.6. Since the normalized adjacency matrix, $\hat{\mathbf{A}}$, is nonnegative by its definition, it will satisfy all the properties that are stated in the Perron–Frobenius Theorem for nonnegative matrices. In other words, it has an algebraically simple eigenvalue which is equal to the spectral radius $\rho(\hat{\mathbf{A}})$ and the corresponding eigenvector is a positive vector. Also, from the graph theory and using the Perron–Frobenius Theorem, we know that $\rho(\hat{\mathbf{A}}) = 1$ for a normalized adjacency matrix. Hence, 1 is an eigenvalue of $\hat{\mathbf{A}}$ and the corresponding eigenvector will have positive entries. This applies to both the right and the left eigenvectors of $\hat{\mathbf{A}}$. Using the relationship between the normalized Laplacian matrix $\hat{L}$ and the normalized adjacency matrix $\hat{\mathbf{A}}$ [34], i.e., $\hat{L} = I - \hat{\mathbf{A}}$, the zero eigenvalue of $\hat{L}$ corresponds to the 1 eigenvalue of $\hat{\mathbf{A}}$, and the corresponding left and right eigenvectors of $\hat{L}$ are the same as those of $\hat{A}$. The right eigenvector is $\mathbf{1}$ but the left eigenvector is not $\mathbf{1}$ for directed graphs in general, unless they are balanced. However, this vector has entries with the same sign. This completes the proof of this lemma. □

Proof of Theorem 4.3. First, we should show that if the control laws defined in (3.31)–(3.33) are modified according to the switching control laws given by (4.20)–(4.22), the matrix $\bar{L}_\sigma$ can be transformed into a balanced matrix. Then, we will use this property to find a common Lyapunov function for the overall switching system. Using the results that are obtained in Lemma 3.2, we know that the following relationships hold between K^i (K_σ^i) and Q^{ij} (Q_σ^i) for an infinite horizon problem:

$$\begin{aligned} &2|N^i|Q^{ij} - \frac{1}{2}K^i(R^i)^{-1}K^i = 0, \; i = 2, \ldots, N, \\ &2(|N^1|Q^{1j} + \Gamma) - \frac{1}{2}K^1(R^1)^{-1}K^1 = 0. \end{aligned} \tag{A.33}$$

For sake of notational simplicity let us assume that all the design parameter matrices are diagonal as follows:

$$Q^{ij} = q^i I, \; R^i = r^i I, \; \Gamma = \gamma I, \tag{A.34}$$

where q^i, r^i, and γ are positive scalars. Then the solutions to (A.33) are given by

$$K^1 = 2\sqrt{(|N^1|q^1+\gamma)r^1}I, \quad K^i = 2\sqrt{|N^i|q^i r^i}I, \ i=2,\dots,N. \tag{A.35}$$

We have seen in Sect. 4.2.1 that in order to make the matrix $\bar{L}_\sigma$ balanced, (4.19) should be satisfied. This implies that the optimal design parameters $Q^{ij}(Q^i_\sigma), R^i(R^i_\sigma)$, and $\Gamma(\Gamma_\sigma)$ should be selected appropriately so that (4.19) is guaranteed.

Let us denote the ith element of the vector ω_σ as $\rho_{i,\sigma}$. Using the definition of μ_σ given in (4.18), we should have

$$\begin{aligned}
\mu_\sigma^T &= ((K^1_\sigma)^{-1}Q^1_\sigma l_{11} \ (K^2_\sigma)^{-1}Q^2_\sigma l_{22} \ \dots \ (K^N_\sigma)^{-1}Q^N_\sigma l_{NN}) \\
&= \left(\frac{|N^1|q^1}{2\sqrt{(|N^1|q^1+\gamma)r^1}} \ \frac{|N^2|q^2}{2\sqrt{|N^2|q^2r^2}} \ \cdots \ \frac{|N^N|q^N}{2\sqrt{|N^N|q^Nr^N}} \right) \otimes I_n \\
&= \kappa \left[\rho_{1,\sigma} \ \rho_{2,\sigma} \ \dots \ \rho_{N,\sigma} \right] \otimes I_n.
\end{aligned} \tag{A.36}$$

Therefore, the following relationships should be satisfied:

$$\kappa\rho_{1,\sigma} = \frac{|N^1|q^1}{2\sqrt{(|N^1|q^1+\gamma)r^1}}, \quad \kappa\rho_{i,\sigma} = \frac{|N^i|q^i}{2\sqrt{|N^i|q^i r^i}}, \ i=2,\dots,N. \tag{A.37}$$

In the first equation of (A.37), $\rho_{1,\sigma}$ and $|N^1|$ are given and κ, q^1, r^1, γ are parameters to be selected. Similarly, in the second equation of (A.37), κ, q^i, r^i are to be selected. It is assumed that r^i and γ are set to fixed values and one then tries to determine q^i that satisfies the above equations. Design of κ is discussed in Sect. 4.2.2. Therefore, the following equations in terms of q^i should be satisfied:

$$|N^1|^2(q^1)^2 - (4\kappa^2\rho_{1,\sigma}^2|N^1|r^1)q^1 - 4(\kappa\rho_{1,\sigma})^2\gamma r^1 = 0, \ q^i = \frac{4(\kappa\rho_{i,\sigma})^2 r^i}{|N^i|}, \ i=2,\dots,N. \tag{A.38}$$

It is not difficult to show that the first equation in (A.38) always has a positive solution $q^1 = \frac{2\kappa\rho_{1,\sigma}}{|N^1|}(\kappa\rho_{1,\sigma}r^1 + \sqrt{(\kappa\rho_{1,\sigma}r^1)^2+\gamma r^1})$. Also, from the second equation of (A.38), it is obvious that $q^i, i=2,\dots,N$ is always positive.

It should be noted that for the above results to hold one should ensure a property in the left null space of $\hat{L}_\sigma$. Namely, due to the positive definiteness of $(K^i_\sigma)^{-1}Q^i_\sigma l_{ii}$, all the elements of the vector μ_σ are of the same sign, i.e., positive, which implies that the null space of $\hat{L}_\sigma$ should also enjoy this property. This can be shown by using the results that are provided in Lemma 4.6.

We are now in a position to use the above relationships to determine the switching control law. From Lemma 3.2, the control inputs can be calculated by using (3.31)–(3.33). By replacing q^i from (A.38) and K^i from (A.35) we obtain

$$\Gamma^1 = -2\frac{|N^1|q^1}{2\sqrt{(|N^1|q^1+\gamma)r^1}}I_n = -2\kappa\rho_{1,\sigma}I_n,$$

$$\Gamma^i = -2\frac{|N^i|q^i}{2\sqrt{|N^i|q^i r^i}}I_n = -2\kappa\rho_{i,\sigma}I_n, \ \ i=2,\dots,N,$$

$$\beta^1 = -2\frac{\gamma}{2\sqrt{(|N^1|q^1+\gamma)r^1}}I_n = -\gamma\left(\kappa\rho_{1,\sigma}r^1 + \sqrt{(\kappa\rho_{1,\sigma}r^1)^2+\gamma r^1}\right)I_n, \quad \text{(A.39)}$$

where $\rho_{i,\sigma}$ is found from the Laplacian matrix of the network at each switching stage. The control laws provided in (4.20)–(4.22) with the parameters as in (A.39) guarantee that $\mathbf{1}^T \otimes I_n$ is in the left null space of $\bar{L}_\sigma$, and therefore $\bar{L}_\sigma$ is the Laplacian of a balanced graph.

Now, to show the stability of the closed-loop switching system we should select a common Lyapunov function candidate which is valid for all the switching states. Let us select the Lyapunov function candidate as $V = \frac{1}{2}e^T e$. Its time derivative along the trajectories of (4.24) is given by $\dot{V} = \frac{1}{2}e^T(L_{cl,\sigma}+L_{cl,\sigma}^T)e = -e^T(\bar{L}_\sigma + \bar{L}_\sigma^T + K_\sigma^{-1}G_\sigma + G_\sigma^T K_\sigma^{-1})e$. Based on the previous discussions $\bar{L}_\sigma = K_\sigma^{-1}Q_\sigma L_\sigma \otimes I_n$ can be considered as the Laplacian matrix of a weighted balanced graph, and by using the results provided in [79], $\bar{L}_\sigma + \bar{L}_\sigma^T$ is also a valid Laplacian matrix representing an undirected (due to its symmetry) and connected graph. Hence, it is a *PSD* matrix. Moreover, the second term, i.e., $K_\sigma^{-1}G_\sigma$ is a diagonal matrix with one nonzero element and so is *PSD*. Hence, $L_{cl,\sigma}+L_{cl,\sigma}^T$ is at least *NSD*. Also, similar to the discussion provided in the proof of Theorem 3.2, the null spaces of the two matrices $\bar{L}_\sigma + \bar{L}_\sigma^T$ and $K_\sigma^{-1}G_\sigma$ do not have any common intersection, and hence their summation is a *PD* matrix and hence $\dot{V} < 0$. Consequently, we can guarantee consensus achievement, and therefore the proof is complete. □

Remark A.1. For evaluating the control laws (4.20) and (4.21) at each switching interval, each agent is required to compute an eigenvector of the network Laplacian matrix. In other words, the Laplacian matrix should be known to all the agents, which is guaranteed by Assumption 4.1.

Proof of Lemma 4.7. Similar to the proof of Theorem 4.3 and without loss of generality, assume that all the matrices involved are diagonal matrices. For the followers' case, we will then have $\frac{\lambda_{\max}(Q^{ij})}{\lambda_{\max}(R^i)} = \frac{q^i}{r^i} = \frac{4(\kappa\rho_{i,\sigma})^2}{|N^i|} > m_i, \ i=2,\dots,N$, and given that $\forall i, \ \rho_{i,\sigma} \neq 0$ (Lemma 4.6), we have $\kappa^2 > \frac{\max_{i=2,\dots,N}(m_i|N^i|)}{4\min_{i=2,\dots,N}(\rho_{i,\sigma}^2)}$. On the other hand for the leader agent, we have the following relationship:

$$q^1 = \frac{2\kappa\rho_{1,\sigma}}{|N^1|}\left(\kappa\rho_{1,\sigma}r^1 + \sqrt{(\kappa\rho_{1,\sigma}r^1)^2+\gamma r^1}\right) > \frac{4\kappa^2\rho_{1,\sigma}^2 r^1}{|N^1|}. \quad \text{(A.40)}$$

To satisfy the condition $\frac{q^1}{r^1} > m_1$, it is sufficient to select κ so that $\kappa^2 > \frac{m_1|N^1|}{4\rho_{1,\sigma}^2}$. Consequently, κ should satisfy the following inequality, namely

$$\kappa^2 > \frac{1}{4}\max\left\{\frac{m_1|N^1|}{\rho_{1,\sigma}^2}, \frac{\max_{i=2,\ldots,N}(m_i|N^i|)}{\min_{i=2,\ldots,N}(\rho_{i,\sigma}^2)}\right\}. \tag{A.41}$$

□

A.3 Proofs of the Lemmas and Theorems of Chap. 5

Proof of Theorem 5.1. (a) Follows from the constructive results that are derived in Sects. 5.1.2 and 5.1.3.

(b) For stability analysis of the closed-loop system we should note that condition 3 in (5.10) guarantees that matrix P has at least the same zeros as the Laplacian matrix of the information graph, L (i.e., it may have more zeros too). Also, since both B and R are block diagonal matrices, the term $BR^{-1}B^TP$ has at least the same zero elements as L. Also, based on its definition as given in (2.8), the matrix A has this property as well. Therefore, the closed-loop matrix A_c has a structure similar to the Laplacian of a subgraph of the original graph, but this subgraph may not be in general a connected one (due to the extra zero entries that may appear in A_c). However, the condition given in (5.11) guarantees that A_c does not have any other zero besides the ones that exist in the Laplacian matrix of one of the connected subgraphs of the original graph. Therefore, A_c has the minimum required nonzero elements to describe the Laplacian matrix of a strongly connected graph. Moreover, since A_c has the structure of a Laplacian matrix and it also satisfies condition 2 in (5.10), it is in fact the Laplacian matrix of a weighted and strongly connected graph. From the graph theory and in particular as shown in [79], it is known that the Laplacian matrix of any strongly connected graph has one and only one zero eigenvalue and $N-1$ negative eigenvalues [34]. Therefore, it is a stable matrix with one zero eigenvalue.

(c) Assume that P is obtained from the optimization problem (5.10). Then we have

$$\begin{aligned}\int_0^T \frac{\mathrm{d}}{\mathrm{d}t}(X^TPX)\mathrm{d}t &= \int_0^T [(AX+BU)^TPX + X^TP(AX+BU)]\mathrm{d}t \\ &= X^T(T)PX(T) - X^T(0)PX(0),\end{aligned}$$

so that we have

$$\int_0^T [(AX+BU)^TPX + X^TP(AX+BU)]\mathrm{d}t + X^T(0)PX(0) - X^T(T)PX(T) = 0. \tag{A.42}$$

By adding the above expression to the cost function (5.1) we get

$$\begin{aligned} J^c &= \int_0^T [X^T QX + U^T RU + (AX+BU)^T PX + X^T P(AX+BU)]\mathrm{d}t \\ &\quad + X^T(0)PX(0) - X^T(T)PX(T), \\ &= \int_0^T [X^T(Q + A^T P + PA)X + U^T RU + U^T B^T PX + X^T PBU]\mathrm{d}t \\ &\quad + X^T(0)PX(0) - X^T(T)PX(T). \end{aligned}$$

Since P is a solution to (5.10), it satisfies (5.5) as well and therefore one gets

$$PA + A^T P - PBR^{-1}B^T P + Q \geq 0. \tag{A.43}$$

Hence,

$$\begin{aligned} J^c &\geq \int_0^T [X^T(PBR^{-1}B^T P)X + U^T RU + U^T B^T PX + X^T PBU]\mathrm{d}t \\ &\quad + X^T(0)PX(0) - X^T(T)PX(T) \\ &= \int_0^T [(U + R^{-1}B^T PX)^T R(U + R^{-1}B^T PX)]\mathrm{d}t \\ &\quad + X^T(0)PX(0) - X^T(T)PX(T). \end{aligned}$$

In order to minimize the integral part of the above cost function one may select the control input as $U^* = -R^{-1}B^T PX$, which is already satisfied due to the definition of the control law.

Moreover, since P is obtained through the optimization problem (5.10), then it is guaranteed that in steady state consensus will be achieved. In other words, if we assume that T is sufficiently large (or $T \to \infty$) to let the system reach a steady state, then $X(T) = \xi[1\ 1\ \ldots\ 1]^T$. Correspondingly, it can be shown that $X^T(T)PX(T) = \xi^2 \sum_i \sum_j P(i,j)$, where $P(i,j)$ represents the ijth entry of the matrix P. Therefore, when $T \to \infty$ the optimal cost has a lower bound that is given by

$$J^{c*} \geq X^T(0)PX(0) - \xi^2 \sum_i \sum_j P(i,j). \tag{A.44}$$

Therefore, $X^T(0)PX(0) - \xi^2 \sum_i \sum_j P(i,j)$ is the finite infimum of J^c. □

Proof of Lemma 5.2. An optimal stabilizing solution for the minimization problem (5.27), or (5.35), exists if the pair $(\bar{A}, \bar{B})$ is stabilizable and the pair $(\bar{A}, \hat{Q})$ (or $(\bar{A}, \Omega)$, $\hat{Q} = \Omega\Omega^*$) is detectable [3]. Each of these conditions can be checked through an *LMI* [17]:

1. $(\bar{A}, \bar{B})$ stabilizable $\Leftrightarrow \bar{A}P_1 + P_1\bar{A}^* < \bar{B}\bar{B}^*$ has a *PD* solution for P_1.

2. $(\bar{A},\hat{Q})$ detectable $\Leftrightarrow \bar{A}^*P_1 + P_1\bar{A} < \hat{Q}$ has a *PD* solution for P_1.

Let us define a new variable $P_2 > 0$ such that

$$P_2 = \bar{S}\bar{S}^*P_2\bar{S}\bar{S}^* + SS^*P_2SS^*. \tag{A.45}$$

An example of such a matrix is the following:

$$P_2 = [\bar{S}\ S]\begin{bmatrix} P_1 & 0 \\ 0 & M_1 \end{bmatrix}\begin{bmatrix} \bar{S}^* \\ S^* \end{bmatrix},\ M_1 > 0. \tag{A.46}$$

Now, we have $P_2\bar{S} = \bar{S}P_1$. By using the definition of the matrices $\bar{A},\bar{B}$, the following conditions should be satisfied to verify stabilizability and detectability conditions, namely

$$\begin{aligned}
&(1)\ \bar{A}P_1 + P_1\bar{A}^* < \bar{B}\bar{B}^* \ \Leftrightarrow\ \bar{S}^*A\bar{S}P_1 + P_1\bar{S}^*A^*\bar{S} < \bar{S}^*BB^*\bar{S} \\
&\Leftrightarrow \bar{S}^*(AP_2 + P_2A^*)\bar{S} < \bar{S}^*BB^*\bar{S} \Leftrightarrow\ \bar{S}^*(AP_2 + P_2A^* - BB^*)\bar{S} < 0 \\
&(2)\ P_1\bar{A} + \bar{A}^*P_1 < \hat{Q}\ \Leftrightarrow P_1\bar{S}^*A\bar{S} + \bar{S}^*A^*\bar{S}P_1\ < \bar{S}^*Q\bar{S} \\
&\Leftrightarrow \bar{S}^*(P_2A + A^*P_2)\bar{S} < \bar{S}^*Q\bar{S}\ \Leftrightarrow \bar{S}^*(P_2A + A^*P_2 - Q)\bar{S} < 0
\end{aligned} \tag{A.47}$$

which are the same conditions as the ones given in (5.36) and (5.37). □

Proof of Theorem 5.2. (a) First, note that from the previous discussions, we have $P^{-1} = \bar{S}^*Z\bar{S}$. Since $\text{rank}(\bar{S}) = Nn - 1$ and $Z > 0$, we have $\bar{S}^*Z\bar{S} > 0$, and hence $P = (\bar{S}^*Z\bar{S})^{-1}$. Therefore

$$\min \text{trace}(P) = \min \text{trace}((\bar{S}^*Z\bar{S})^{-1}). \tag{A.48}$$

This in turn can be written equivalently by introducing an auxiliary variable Γ such that [45]

$$\min \text{trace}(\Gamma)\ \text{s.t.}\ \begin{bmatrix} \Gamma & I \\ I & \bar{S}^*Z\bar{S} \end{bmatrix} > 0, \tag{A.49}$$

which is the first condition to be considered. Moreover, by twice applying the Schur complement to (5.33), the second inequality follows. Furthermore, by using the definition of the matrix Z [24], it can be shown that Condition 3 is equivalent to the consensus constraint given by (5.15). Finally, the last condition is an assumption on the structure of the new variable Z which has already been discussed.

In summary, the minimization formulation, together with the first two inequalities in (5.38) and condition 4, is used to design an optimal stabilizing controller for the system (5.17) as discussed previously. The third equality is used to guarantee that the consensus is achieved.

(b) The matrix $K = WZ^{-1}$ will have the same structure as W if Z is selected to be diagonal [24]. Therefore, we may transform any required constraint on the control gain matrix to that of the matrix W by considering Z to be diagonal. Therefore, if the individual controllers are to be designed based on information received from the neighbors of each agent, the structure for W may be chosen as the Laplacian matrix, so that the members' information in the neighboring sets are sufficient for design of each agent's control signal. □

Proof of Lemma 5.3. Based on the definition of the matrices A, B given by (2.8) and the restriction on the structure of K as provided in Theorem 5.2, the matrix $A+BK$ has a structure similar to that of the Laplacian matrix of the entire network, with the possibility of some additional zero elements. This matrix should also satisfy the condition in (5.15), and hence it can be considered as a Laplacian matrix [34]. Moreover, the graph corresponding to this matrix has at most the same edges as the graph of the entire network. Therefore, it represents a subgraph of the network graph with different edges' weights. □

Proof of Theorem 5.3. (a) As discussed in Sect. 5.2.1, two conditions should be satisfied to guarantee consensus achievement. One of these conditions is (5.15) and the other one is the design of matrix K to ensure that the system in (5.17) is asymptotically stable. In the following, we show that connectedness of the underlying network graph is necessary for satisfying the latter condition.

To satisfy the asymptotic stability condition, the matrix $\bar{S}^*(A+BK)\bar{S}$ in (5.17) should be Hurwitz. In other words, it should have no zero eigenvalue. Now, we show that this is violated if the network graph is not connected. Based on its definition, matrix $\bar{S}_{Nn\times(Nn-1)}$ consists of $Nn-1$ independent column vectors. Denote these vectors by $\bar{S}_1, \ldots, \bar{S}_{Nn-1}$, i.e.

$$\bar{S} = [\bar{S}_1,\ \bar{S}_2,\ \ldots, \bar{S}_{Nn-1}]. \tag{A.50}$$

Then, we will have

$$\begin{aligned}\bar{S}^*(A+BK)\bar{S} &= [\bar{S}_1,\ \bar{S}_2,\ \ldots, \bar{S}_{Nn-1}]^*(A+BK)[\bar{S}_1,\ \bar{S}_2,\ \ldots, \bar{S}_{Nn-1}] \\ &= \begin{bmatrix} \bar{S}_1^*(A+BK)\bar{S}_1 & \ldots\ \bar{S}_1^*(A+BK)\bar{S}_{Nn-1} \\ \ddots & \\ \bar{S}_{Nn-1}^*(A+BK)\bar{S}_1 & \ldots\ \bar{S}_{Nn-1}^*(A+BK)\bar{S}_{Nn-1} \end{bmatrix}.\end{aligned}$$

Now, we assume that the network underlying graph is not connected. Then, the Laplacian matrix of the graph L may have more than one zero eigenvalue. Correspondingly, $A+BK$ represents a disconnected graph and therefore has more than one zero eigenvalue. Hence, $A+BK$ has an eigenvector corresponding to one of its zero eigenvalues which is not necessarily in the $\mathscr{S}$ subspace. Let us denote this eigenvector by w, i.e., $(A+BK)w = 0$. This vector in general may have components in both subspaces $\mathscr{S}$ and $\bar{\mathscr{S}}$, i.e.,

$$w = \alpha_1 \bar{S}_1 + \alpha_2 \bar{S}_2 + \cdots + \alpha_{(Nn-1)} \bar{S}_{Nn-1} + \gamma S, \tag{A.51}$$

with at least one nonzero α_i. Now, assume that α_1 is nonzero and since $(A+BK)S = 0$ and $(A+BK)w = 0$ we will have

$$\begin{aligned}
&(A+BK)(\alpha_1 \bar{S}_1 + \alpha_2 \bar{S}_2 + \cdots + \alpha_{Nn-1} \bar{S}_{Nn-1} + \gamma S) = 0 \\
&\Rightarrow (A+BK)(\alpha_1 \bar{S}_1 + \alpha_2 \bar{S}_2 + \cdots + \alpha_{Nn-1} \bar{S}_{Nn-1}) = 0 \\
&\Rightarrow (A+BK)\bar{S}_1 = \frac{1}{\alpha_1}(-\alpha_2 (A+BK)\bar{S}_2 - \cdots - \alpha_{Nn-1}(A+BK)\bar{S}_{Nn-1}) \\
&\Rightarrow \begin{cases}
\bar{S}_1^*(A+BK)\bar{S}_1 = \frac{1}{\alpha_1}(-\alpha_2 \bar{S}_1^*(A+BK)\bar{S}_2 - \cdots - \alpha_{Nn-1}\bar{S}_1^*(A+BK)\bar{S}_{Nn-1}) \\
\bar{S}_2^*(A+BK)\bar{S}_1 = \frac{1}{\alpha_1}(-\alpha_2 \bar{S}_2^*(A+BK)\bar{S}_2 - \cdots - \alpha_{Nn-1}\bar{S}_2^*(A+BK)\bar{S}_{Nn-1}) \\
\vdots \\
\bar{S}_{Nn-1}^*(A+BK)\bar{S}_1 = \\
\frac{1}{\alpha_1}(-\alpha_2 \bar{S}_{Nn-1}^*(A+BK)\bar{S}_2 - \cdots - \alpha_{Nn-1}\bar{S}_{Nn-1}^*(A+BK)\bar{S}_{Nn-1}).
\end{cases}
\end{aligned}$$

If at least one of the α_i's besides α_1 is nonzero, then from the above equations we can conclude that at least one column of the matrix $\bar{S}^*(A+BK)\bar{S}$ can be written as a linear combination of other columns, i.e., $\bar{S}^*(A+BK)\bar{S}$ is rank deficient and so has a zero eigenvalue. This is contradictory with the asymptotic stability condition of system (5.17). On the other hand, if all the α_i's are zero except α_1, then $w = \alpha_1 \bar{S}_1 + \gamma S$.

Therefore, $(A+BK)(\alpha_1 \bar{S}_1 + \gamma S) = 0$ and hence $(A+BK)\bar{S}_1 = 0$. This implies that the first column of the matrix $\bar{S}^*(A+BK)\bar{S}$ is zero, which results in singularity of this matrix. Therefore, we may conclude that if the underlying graph is not connected, the asymptotic stability of system (5.17) cannot be guaranteed. This implies that consensus achievement cannot be guaranteed. In other words, connectivity of the underlying network graph is a necessary condition for the consensus achievement.

(b) We have previously shown that $A+BK$ should have the same structure as the L matrix with some possible additional zero elements. Therefore, it describes a subgraph of the original graph. On the other hand, based on the discussions in the previous part in order to guarantee consensus, $A+BK$ should describe the Laplacian of a connected graph. This implies that if a matrix K satisfies the *LMI* conditions provided in (5.38), then the additional zero elements are such that $A+BK$ represents a connected network. Therefore, $A+BK$ describes the Laplacian matrix of a connected subgraph of the original connected underlying network graph. □

References

1. M. Ait-Rami and X. Y. Zhou, *Linear Matrix Inequalities, Riccati equations, and indefinite stochastic linear quadratic controls*, IEEE Trans. Autom. Control **45** (2000), no. 6, 1131–1142.
2. M. Alighanbari and J.P. How, *Decentralized task assignment for unmanned aerial vehicles*, Proc. Conference on Decision and Control and European Control Conference, December 12–15, 2005, pp. 5668–5673.
3. B. D. O. Anderson and J. B. Moore, *Optimal control: Linear quadratic methods*, Prentice Hall Information and System Sciences Series, 1990.
4. B. Andrea-Novel, G. Bastin, and G. Campion, *Modeling and control of non holonomic wheeled mobile robots*, Proc. international conference on robotics and automation, 1991, pp. 1130–1135.
5. D. Angeli and P.A. Bliman, *Tight estimates for non-stationary consensus with fixed underlying spanning tree*, Proc. IFAC world congress, 2008, pp. 9021–9026.
6. M. Arcak, *Passivity as a design tool for group coordination*, Proc. American Control Conference, June 14–16, 2006, pp. 29–34.
7. ______, *Passivity as a design tool for group coordination*, IEEE Trans. Autom. Control **52** (2007), no. 8, 1380–1390.
8. T. Baleh and R.C. Arkin, *Behavior-based formation control for multi-robot teams*, IEEE J. Robot. Autom. **14** (1998), 926–939.
9. T. Basar and G.J. Olsder, *Dynamic noncooperative game theory*, Academic Press Inc., 1982.
10. D. Bauso, L. Giarre, and R. Pesenti, *Mechanism design for optimal consensus problems*, Proc. Conference on Decision and Control, December 13–15, 2006, pp. 3381–3386.
11. ______, *Robust control in uncertain multi-inventory systems and consensus problems*, Proc. IFAC world congress, 2008, pp. 9027–9032.
12. R. W. Beard, J. Lawton, and F. Y. Hadaegh, *A feedback architecture for formation control*, Proc. American Control Conference, vol. 6, June 28–30, 2000, pp. 4087–4091.
13. ______, *A coordination architecture for spacecraft formation control*, IEEE Trans. Control Syst. Technol. **9** (2001), no. 6, 777–789.
14. P. Benner and H. Mena, *BDF methods for large-scale differential equations*, Proc. 16th international symposium on mathematical theory of network and systems (MTNS), July 5–9, 2004.
15. J. D. Bošković, S. E. Bergstrom, and R. K. Mehra, *Retrofit reconfigurable flight control in the presence of control effector damage*, Proc. American Control Conference, June 8–10, 2005, pp. 2652–2657.
16. J. D. Bošković, S. M. Li, and R. K. Mehra, *Formation flight control design in the presence of unknown leader commands*, Proc. American Control Conference, May 8–10, 2002, pp. 2854–2859.

E. Semsar-Kazerooni and K. Khorasani, *Team Cooperation in a Network of Multi-Vehicle Unmanned Systems: Synthesis of Consensus Algorithms*, DOI 10.1007/978-1-4614-5073-3, © Springer Science+Business Media New York 2013

17. S. Boyd, L. El Ghaoui, E. Feron, and V. Balakrishnan, *Linear Matrix inequalities in system and control theory*, SIAM, 1994.
18. W. L. Brogan, *Modern control theory*, Prentice Hall, 1991.
19. F. Bullo, J. Cortés, and S. Martinez, *Distributed control of robotic networks*, series in applied mathematics ed., Princeton, 2009.
20. G. Campion, G. Bastin, and B. D Andrea-Novel, *Structural properties and classification of kinematic and dynamic models of wheeled mobile robots*, IEEE Trans. Robot. Autom. **12** (1996), no. 1, 47–62.
21. M. Cao, D.A. Spielman, and A.S. Morse, *A lower bound on convergence of a distributed network consensus algorithm*, Proc. Conference on Decision and Control, 2005, pp. 2356–2361.
22. S. Chaiken, *A combinatorial proof of the all minors matrix tree theorem*, SIAM J. Alg. Disc. Meth. **3** (1982), no. 3, 319–329.
23. J. Cortes, S. Martinez, and F. Bullo, *Robust rendezvous for mobile autonomous agents via proximity graphs in arbitrary dimensions*, IEEE Trans. Autom. Control **52** (2003), 166–180.
24. G. A. Decastro and F. Paganini, *Convex synthesis of controllers for consensus*, Proc. American Control Conference, June 30–July 2, 2004, pp. 4933–4938.
25. J. C. Delvenne, R. Carli, and S. Zampieri, *Optimal strategies in the average consensus problem*, Proc. Conference on Decision and Control-European Control Conference (CDC-ECC), 12–14, 2007, pp. 2498–2503.
26. J. C. Engwerda, *LQ dynamic optimization and differential games*, John Wiley & Sons, 2005.
27. F. Fagnani and S. Zampieri, *Asymmetric randomized gossip algorithms for consensus*, Proc. IFAC world congress, 2008, pp. 9051–9056.
28. L. Fang, P. J. Antsaklis, and A. Tzimas, *Asynchronous consensus protocols: Preliminary results, simulations, and open questions*, Proc. Conference on Decision and Control-European Control Conference (CDC-ECC), 12–15, 2005, pp. 2194–2199.
29. J. A. Fax, *Optimal and cooperative control of vehicle formations*, Ph.D. thesis, California Institute of Technology, 2002.
30. J. A. Fax and R. M. Murray, *Information flow and cooperative control of vehicle formations*, IEEE Trans. Autom. Control **49** (2004), no. 9, 1465–1476.
31. R. Fierro and A. K. Das, *A modular architecture for formation control*, Proc. IEEE workshop on robot motion and control, November 9–11, 2002, pp. 285–290.
32. V. Gazi, *Stability analysis of swarms*, Ph.D. thesis, The Ohio State University, 2002.
33. ______, *Stability of an asynchronous swarm with time-dependent communication links*, IEEE Trans. Syst., Man, Cybern. B **38** (2008), no. 1, 267–274.
34. C. Godsil and G. Royle, *Algebraic graph theory*, Springer, 2001.
35. V. Gupta, B. Hassibi, and R. M. Murray, *On the synthesis of control laws for a network of autonomous agents*, Proc. American Control Conference, vol. 6, June 30–July 2, 2004, pp. 4927–4932.
36. F. Y. Hadaegh, A.R. Ghavimi, G. Singh, and M. Quadrelli, *A centralized optimal controller for formation flying spacecraft*, Proc. Int. conf. intel. tech., 2000.
37. F. Y. Hadaegh, D.P. Scharf, and S.R. Ploen, *Initialization of distributed spacecraft for precision formation flying*, Proc. Conference on Control Applications, vol. 2, June 23–25, 2003, pp. 1463–1468.
38. S. Hirche and S. Hara, *Stabilizing interconnection characterization for multi-agent systems with dissipative properties*, Proc. IFAC world congress, 2008, pp. 1571–1577.
39. Y. C. Ho and K. C. Chu, *Team decision theory and information structures in optimal control problems-part I*, IEEE Trans. Autom. Control **17** (1972), no. 1, 15–22.
40. ______, *Team decision theory and information structures in optimal control problems-part II*, IEEE Trans. Autom. Control **17** (1972), no. 1, 22–28.
41. R. A. Horn and C. R. Johnson, *Matrix analysis*, Cambridge University Press, 1990.
42. Q. Hui and W. M. Haddad, *H_2 optimal semistable stabilization for linear discrete-time dynamical systems with applications to network consensus*, Proc. Conference on Decision and Control (CDC), 12–14, 2007, pp. 2315–2320.

43. I.I. Hussein and A.M. Bloch, *Dynamic coverage optimal control for interferometric imaging spacecraft formations (part ii): the nonlinear case*, Proc. American Control Conference, June 8–10, 2005, pp. 2391–2396.
44. G. İnalhan, D. M. Stipanović, and C. J. Tomlin, *Decentralized optimization, with application to multiple aircraft coordination*, Proc. Conference on Decision and Control, December 10–13, 2002, pp. 1147–1155.
45. A. Jadbabaie, *Robust, non-fragile controller synthesis using model-based fuzzy systems: A linear matrix inequality approach*, Master's thesis, The University of New Mexico, 1997.
46. A. Jadbabaie, J. Lin, and S. Morse, *Coordination of groups of mobile autonomous agents using nearest neighbor rules*, IEEE Trans. Autom. Control **48** (2003), no. 6, 988–1000.
47. S. Jinyan, W. Long, and Y. Junzhi, *Cooperative control of multiple robotic fish in a disk-pushing task*, Proc. American Control Conference, June 14–16, 2006, pp. 2730–2735.
48. H. K. Khalil, *Nonlinear systems*, Prentice Hall, 1996.
49. D. B. Kingston and R. W. Beard, *Discrete-time average-consensus under switching network topologies*, Proc. American Control Conference, June 14–16, 2006, pp. 3551–3556.
50. T. J. Koo and S. M. Shahruz, *Formation of a group of unmanned aerial vehicles (UAVs)*, Proc. American Control Conference, June 25–27, 2001, pp. 69–74.
51. J. Lawton, R. W. Beard, and F. Y. Hadaegh, *An adaptive control approach to satellite formation flying with relative distance constraints*, Proc. American Control Conference, vol. 3, June 2–4, 1999, pp. 1545–1549.
52. D. Lee and M. W. spong, *Agreement with non-uniform information delays*, Proc. American Control Conference, June 14–16, 2006, pp. 756–761.
53. ______, *Stable flocking of multiple inertial agents on balanced graphs*, Proc. American Control Conference, June 14–16, 2006, pp. 2136–2141.
54. ______, *Stable flocking of inertial agents on balanced communication graphs*, IEEE Trans. Autom. Control **52** (2007), no. 8, 1469–1475.
55. N.E. Leonard and E. Fiorelli, *Virtual leaders, artificial potentials and coordinated control of groups*, Proc. Conference on Decision and Control, June 30–July 2, 2001, pp. 2968–2973.
56. T. Li, *Asymptotically unbiased average consensus under measurement noises and fixed topologies*, Proc. IFAC world congress, 2008, pp. 2867–2873.
57. B. Liu, G. Xie, T. Chu, and L. Wang, *Controllability of interconnected systems via switching networks with a leader*, Proc. IEEE Conference on Systems, Man, and Cybernetics, October 8–11, 2006, pp. 3912–2849.
58. A. Locatelli, *Optimal control: An introduction*, Birkhauser, 2001.
59. H. Lutkepohl, *Handbook of matrices*, Wiley publication, 1996.
60. C.Q. Ma, T. Li, and J.F. Zhang, *Leader-following consensus control for multi-agent systems under measurement noises*, Proc. IFAC world congress, 2008, pp. 1528–1533.
61. J. Marschak, *Elements of a theory of teams*, Management Science **1** (1955), 127–137.
62. J. Marschak and R. Radner, *Economic theory of teams*, Yale University Press, 1972.
63. M. Mesbahi and F. Y. Hadaegh, *Reconfigurable control for the formation flying of multiple spacecraft*, Proc. International Multi-Conference on Circuits, Systems, and Control, 1999.
64. ______, *Formation flying of multiple spacecraft via graphs, matrix inequalities, and switching*, AIAA Journal of Guidance, Control, and Dynamics **24** (2001), no. 2, 369–377.
65. L. Moreau, *Stability of continuous time distributed consensus algorithms*, Proc. Conference on Decision and Control, December 14–17, 2004, pp. 3998–4003.
66. ______, *Stability of multiagent systems with time-dependent communication links*, IEEE Trans. Autom. Control **50** (2005), no. 2, 169–182.
67. N. Moshtagh, A. Jadbabaie, and K. Daniilidis, *Distributed geodesic control laws for flocking of nonholonomic agents*, Proc. Conference on Decision and Control, December 12–15, 2005, pp. 2835–2840.
68. U. Münz, A. Papachristodoulou, and F. Allgöwer, *Nonlinear multi-agent system consensus with time-varying delays*, Proc. IFAC world congress, 2008, pp. 1522–1527.
69. E. Nett and S. Schemmer, *Reliable real-time communication in cooperative mobile applications*, IEEE Trans. Comput. **52** (2003), 166–180.

70. R. Olfati-Saber, *Ultrafast consensus in small-world networks*, Proc. American Control Conference, June 8–10, 2005, pp. 2371–2378.
71. ______, *Flocking for multi-agent dynamic systems: Algorithms and theory*, IEEE Trans. Autom. Control **51** (2006), no. 3, 401–420.
72. R. Olfati-Saber, J. A. Fax, and R. M. Murray, *Consensus and cooperation in networked multi-agent systems*, Proc. of the IEEE **95** (2007), no. 1, 215–233.
73. R. Olfati-Saber and R. M. Murray, *Distributed cooperative control of multiple vehicle formations using structural potential functions*, Proc. IFAC World Congress, 2002.
74. ______, *Distributed structural stabilization and tracking for formations of dynamic multi-agents*, Proc. Conference on Decision and Control, December 10–13, 2002, pp. 209–215.
75. ______, *Graph rigidity and distributed formation stabilization of multi-vehicle systems*, Proc. Conference on Decision and Control, December 10–13, 2002, pp. 2965–2971.
76. ______, *Agreement problems in networks with directed graphs and switching topology*, Proc. Conference on Decision and Control, December 9–12, 2003, pp. 4126–4132.
77. ______, *Consensus protocols for networks of dynamic agents*, Proc. American Control Conference, June 4–6, 2003, pp. 951–956.
78. ______, *Flocking with obstacle avoidance: cooperation with limited communication in mobile networks*, Proc. Conference on Decision and Control, December 9–12, 2003, pp. 2022–2028.
79. ______, *Consensus problems in networks of agents with switching topology and time-delays*, IEEE Trans. Autom. Control **49** (2004), no. 9, 1520–1533.
80. D. Paley, N. E. Leonard, and R. Sepulchre, *Collective motion: Bistability and trajectory tracking*, Proc. Conference on Decision and Control, December 14–17, 2004, pp. 1932–1937.
81. R. Radner, *Team decision problems*, Annals of Mathematical Statistics **33** (1962), no. 3, 857–881.
82. R. L. Raffard, C. J. Tomlin, and S. P. Boyd, *Distributed optimization for cooperative agents: application to formation flight*, Proc. Conference on Decision and Control, December 14–17, 2004, pp. 2453–2459.
83. A. Rahmani and M. Mesbahi, *On the controlled agreement problem*, Proc. American Control Conference, June 14–16, 2006, pp. 1376–1381.
84. ______, *A graph-theoretic outlook on the controllability of the agreement dynamics*, Proc. European Control Conference, July 2–5, 2007.
85. ______, *Pulling the strings on agreement: Anchoring, controllability, and graph automorphisms*, Proc. American Control Conference, July 9–13, 2007, pp. 2738–2743.
86. W. Ren, *Distributed attitude consensus among multiple networked spacecraft*, Proc. American Control Conference, 2006, pp. 1760–1765.
87. ______, *Consensus strategies for cooperative control of vehicle formations*, IET Control Theory Applications **1** (2007), no. 2, 505–512.
88. ______, *On consensus algorithms for double-integrator dynamics*, Proc. Conference on Decision and Control, December 12–14, 2007, pp. 2295–2300.
89. ______, *Second-order consensus algorithm with extensions to switching topologies and reference models*, Proc. American Control Conference, July 11–13, 2007, pp. 1431–1436.
90. W. Ren and R. W. Beard, *Consensus of information under dynamically changing interaction topologies*, Proc. American Control Conference, June 30–July 2, 2004, pp. 4939–4944.
91. ______, *Decentralized scheme for spacecraft formation flying via the virtual structure approach*, Journal of guidance, control and dynamics **127** (2004), no. 1, 73–82.
92. ______, *Formation feedback control for multiple spacecraft via virtual structures*, IEE proc. control theory appl. **151** (2004), no. 3, 357–368.
93. ______, *Consensus seeking in multiagent systems under dynamically changing interaction topologies*, IEEE Trans. Autom. Control **50** (2005), no. 5, 655–661.
94. ______, *Distributed consensus in multi-vehicle cooperative control*, communications and control engineering series ed., Springer-Verlag, London, 2008.
95. W. Ren, R. W. Beard, and E. M. Atkins, *A survey of consensus problems in multi-agent coordination*, Proc. American Control Conference, June 8–10, 2005, pp. 1859–1864.

96. W. Ren and Yongcan Cao, *Distributed coordination of multi-agent networks*, communications and control engineering series ed., Springer-Verlag, London, 2011.
97. C. W. Reynolds, *flocks, herds, and schools: a distributed behavioral model*, Proc. Computer Graphics (ACM SIGGRAPH), 1987, pp. 25–34.
98. S. Samar, S. Boyd, and D. Gorinevsky, *Distributed estimation via dual decomposition*, Proc. European Control Conference, 2007, pp. 1511–1519.
99. D. P. Scharf, F. Y. Hadaegh, and S. R. Ploen, *A survey of spacecraft formation flying guidance and control part II: control*, Proc. American Control Conference, June 30–July 2, 2004, pp. 2976–2985.
100. D.P. Scharf, A.B. Acikmese, S.R. Ploen, and F. Y. Hadaegh, *A direct solution for fuel-optimal reactive collision avoidance of collaborating spacecraft*, Proc. American Control Conference, vol. 2, June 14–16, 2006, pp. 5201–5206.
101. E. Semsar and K. Khorasani, *Adaptive formation control of UAVs in the presence of unknown vortex forces and leader commands*, Proc. American Control Conference, June 14–16, 2006, pp. 3563–3568.
102. ______, *Optimal control and game theoretic approaches to cooperative control of a team of multi-vehicle unmanned systems*, Proc. IEEE International Conference on Networking, Sensing and Control, April 15–17, 2007, pp. 628–633.
103. E. Semsar-Kazerooni and K. Khorasani, *Optimal cooperation in a modified leader-follower team of agents with partial availability of leader command*, Proc. IEEE International Conference on Systems, Man, and Cybernetics, October 7–10, 2007, pp. 234–239.
104. ______, *Optimal performance of a modified leader-follower team of agents with partial availability of leader command and presence of team faults*, Proc. IEEE Conference on Decision and Control, December12–14, 2007, pp. 2491–2497.
105. ______, *Semi-decentralized optimal control of a cooperative team of agents*, Proc. IEEE International Conference on System of Systems Engineering, April 17–19, 2007, pp. 1–7.
106. ______, *Semi-decentralized optimal control technique for a leader-follower team of unmanned systems with partial availability of the leader command*, Proc. IEEE International Conference on Control and Automation, May 30–June 1, 2007, pp. 475–480.
107. ______, *Optimal consensus algorithms for cooperative team of agents subject to partial information*, Automatica **44** (2008), no. 11, 2766–2777.
108. ______, *Switching control of a modified leader-follower team of agents under the leader and network topological changes*, Proc. IFAC world congress, vol. 17, 2008, pp. 1534–1540.
109. ______, *Analysis of actuator faults in a cooperative team consensus of unmanned systems*, Proc. American Control Conference, June 10–12, 2009, pp. 2618–2623.
110. ______, *A game theory approach to multi-agent team cooperation*, Proc. American Control Conference, June 10–12, 2009, pp. 4512–4518.
111. ______, *An LMI approach to optimal consensus seeking in multi-agent systems*, Proc. American Control Conference, June 10–12, 2009, pp. 4519–4524.
112. ______, *Multi-agent team cooperation: A game theory approach*, Automatica **45** (2009), no. 10, 2205–2213.
113. ______, *An optimal cooperation in a team of agents subject to partial information*, International Journal of Control **82** (2009), no. 3, 571–583.
114. ______, *On optimal consensus seeking: An LMI approach*, IEEE Trans. Systems, Man, Cybernetics: part B **40** (2010), no. 2, 540–547.
115. ______, *Team consensus for a network of unmanned vehicles in presence of actuator faults*, IEEE Trans. Control System Technology **18** (2010), no. 5, 1155–1161.
116. ______, *Switching control of a modified leader-follower team of agents under the leader and network topological changes*, IET Control Theory and Applications **5** (2011), no. 12, 1369–1377.
117. H. Shi, L. Wang, and T. Chu, *Coordinated control of multiple interactive dynamical agents with asymmetric coupling pattern and switching topology*, Proc. IEEE/RSJ International Conference on Intelligent Robots and Systems, October 9–15, 2006, pp. 3209–3214.

118. S. N. Singh and M. Pachter, *Adaptive feedback linearizing nonlinear close formation control of UAVs*, Proc. American Control Conference, June 28–30, 2000, pp. 854–858.
119. B. Sinopoli, C. Sharp, L. Schenato, S. Schafferthim, and S. Sastry, *Distributed control applications within sensor networks*, Proc. IEEE **91** (2003), no. 8, 1235–1246.
120. R. S. Smith and F. Y. Hadaegh, *Control strategies for deep space formation flying spacecraft*, Proc. American Control Conference, May 8–10, 2002, pp. 2836–2841.
121. ______, *Parallel estimators and communication in spacecraft formations*, Proc. IFAC World Congress, 2005.
122. ______, *A distributed parallel estimation architecture for cooperative vehicle formation control*, Proc. American Control Conference, June 14–16, 2006, pp. 4219–4224.
123. ______, *Distributed parallel estimation architecture for cooperative vehicle formation control*, Proc. IFAC World Congress, July 14–16, 2006.
124. ______, *Closed-loop dynamics of cooperative vehicle formations with parallel estimators and communication*, IEEE Trans. Autom. Control **52** (2007), no. 8, 1404–1414.
125. ______, *Distributed estimation, communication and control for deep space formations*, IET Control Theory Applications **1** (2007), no. 2, 445–451.
126. J. L. Speyer, I. Seok, and A. Michelin, *Decentralized control based on the value of information in large vehicle arrays*, Proc. American Control Conference, June 11–13, 2008, pp. 5047–5054.
127. J. L. Speyer, N. Zika, and E. Franco, *Determination of the value of information in large vehicle arrays*, Proc. American Control Conference, June 14–16, 2006, pp. 1–7.
128. D. M. Stipanović, G. İnalhan, R. Teo, and C. J. Tomlin, *Decentralized overlapping control of a formation of unmanned aerial vehicles*, Automatica **40** (2004), 1285–1296.
129. Y. G. Sun, L. Wang, and G. Xie, *Average consensus in directed networks of dynamic agents with time-varying communication delays*, Proc. Conference on Decision and Control, December 13–15, 2006, pp. 3393–3398.
130. H. G. Tanner, A. Jadbabaie, and G. J. Pappas, *Stable flocking of mobile agents part I: fixed topology*, Proc. Conference on Decision and Control, December 9–12, 2003, pp. 2010–2015.
131. ______, *Stable flocking of mobile agents part II: Dynamic topology*, Proc. Conference on Decision and Control, December 9–12, 2003, pp. 2016–2021.
132. ______, *Flocking in fixed and switching networks*, IEEE Trans. Autom. Control **52** (2007), no. 5, 863–868.
133. H. G. Tanner, V. Kumar, and G. J. Pappas, *The effect of feedback and feedforward on formation ISS*, Proc. IEEE International Conference on Robotics and Automation, 2002, pp. 3448–3453.
134. H. G. Tanner, G. J. Pappas, and V. Kumar, *Leader-to-formation stability*, IEEE J. Robot. Autom. **20** (2004), no. 3, 443–455.
135. J. N. Tsitsiklis, *Problems in decentralized decision making and computation*, Ph.D. thesis, Massachusetts Institute of Technology, 1984.
136. J. N. Tsitsiklis, D. P. Bertsekas, and M. Athans, *Distributed asynchronous deterministic and stochastic gradient optimization algorithms*, IEEE Trans. Autom. Control **31** (1986), no. 9, 803–812.
137. J.Z. Wang, I. Mareels, and Y. Tan, *Robustness of distributed multi-agent consensus*, Proc. IFAC world congress, 2008, pp. 1510–1515.
138. L. Wang and F. Xiao, *A new approach to consensus problems for discrete-time multiagent systems with time-delays*, Proc. American Control Conference, June 14–16, 2006, pp. 2118–2123.
139. P.K.C. Wang, *Navigation strategies for multiple autonomous mobile robots moving in formation*, IEEE J. Robot. Autom. **8** (1991), 177–195.
140. C. W. Wu, *Agreement and consensus problems in groups of autonomous agents with linear dynamics*, Proc. IEEE International Symposium on Circuits and Systems (ISCAS), May 23–26, 2005, pp. 292–295.
141. F. Xiao and L. Wang, *Consensus problems for high-dimensional multi-agent systems*, IET Control Theory Applications **1** (2007), no. 3, 830–837.

142. L. Xiao and S. Boyd, *Fast linear iterations for distributed averaging*, Systems and control letters **53** (2004), no. 1, 65–78.
143. G. Xie and L. Wang, *Consensus control for a class of networks of dynamic agents: Switching topology*, Proc. American Control Conference, June 14–16, 2006, pp. 1382–1387.
144. H. Yamaguchi, *Adaptive formation control for distributed autonomous mobile robot groups*, Proc. International Conference on Robotics and Automation, April 20–25, 1997, pp. 2300–2305.
145. T. Yoshikawa, *Decomposition of dynamic team decision problems*, IEEE Trans. Autom. Control (1978), 627–632.
146. S. Zampieri, *Trends in networked control systems*, Proc. IFAC world congress, 2008, pp. 2886–2894.
147. G. Zimmermann, *A minimax-condition for the characteristic center of a tree*, Linear and Multilinear Algebra **45** (1998), 161–187.

Index

A
Actuator faults, 30
Adjacency matrix, 28
Algebraic Riccati equation, 32

B
Balanced graph, 28
Bargaining problem, 34, 35

C
Connected graph, 28
Consensus, 4
Consensus algorithms, 10
Cooperative games, 33

F
Flocking, 4
Formation control, 3

H
Hamilton–Jacobi–Bellman equations, 30

I
Interaction terms, 22

L
Laplacian matrix, 28
Leaderless structure, 20
Linear Quadratic Regulator, 31

M
Modified Leader–Follower structure, 21

N
Normalized Laplacian matrix, 28

P
Pareto efficient, 33

S
Schur complement, 29
Sensor network, 1
Spectral Radius, 29
State decomposition, 93
Switching network structure, 68

U
Unmanned system network, 1

E. Semsar-Kazerooni and K. Khorasani, *Team Cooperation in a Network of Multi-Vehicle Unmanned Systems: Synthesis of Consensus Algorithms*, DOI 10.1007/978-1-4614-5073-3, © Springer Science+Business Media New York 2013